FOREIGN DIRECT INVEST
AGGLOMERATION AND EXTE

T0227946

Ashgate Economic Geography Series

Series Editors:
Michael Taylor, Peter Nijkamp, and Tom Leinbach

Innovative and stimulating, this quality series enlivens the field of economic geography and regional development, providing key volumes for academic use across a variety of disciplines. Exploring a broad range of interrelated topics, the series enhances our understanding of the dynamics of modern economies in developed and developing countries, as well as the dynamics of transition economies. It embraces both cutting edge research monographs and strongly themed edited volumes, thus offering significant added value to the field and to the individual topics addressed.

Other titles in the series:

Traditional Food Production and Rural Sustainable Development
A European Challenge
Edited by Teresa de Noronha Vaz, Peter Nijkamp and Jean-Louis Rastoin
ISBN: 978-0-7546-7462-7

Upgrading Clusters and Small Enterprises in Developing Countries
Environmental, Labor, Innovation and Social Issues
Edited by Jose Antonio Puppim de Oliveira
ISBN: 978-0-7546-7297-5

The Moving Frontier
The Changing Geography of Production in Labour-Intensive Industries
Edited by Lois Labrianidis
ISBN: 978-0-7546-7448-1

Network Strategies in Europe
Developing the Future for Transport and ICT
Edited by Maria Giaoutzi and Peter Nijkamp
ISBN: 978-0-7546-7330-9

Tourism and Regional Development
New Pathways
Edited by Maria Giaoutzi and Peter Nijkamp
ISBN: 978-0-7546-4746-1

Foreign Direct Investment, Agglomeration and Externalities
Empirical Evidence from Mexican Manufacturing Industries

JACOB A. JORDAAN
VU University Amsterdam, The Netherlands

Routledge
Taylor & Francis Group

LONDON AND NEW YORK

First published 2009 by Ashgate Publishing

2 Park Square, Milton Park, Abingdon, Oxon OX14 4RN
711 Third Avenue, New York, NY 10017, USA

Routledge is an imprint of the Taylor & Francis Group, an informa business

First issued in paperback 2016

British Library Cataloguing in Publication Data
Jordaan, Jacob A.
 Foreign direct investment, agglomeration and externalities
 : empirical evidence from Mexican manufacturing industries.
 -- (Ashgate economic geography series)
 1. Investments, Foreign--Mexico. 2. Manufacturing
 industries--Location--Mexico. 3. Business enterprises,
 Foreign--Location--Mexico. 4. Industrial clusters--
 Mexico. 5. Mexico--Economic conditions--Regional
 disparities. 6. Externalities (Economics) 7. Strategic
 alliances (Business)--Mexico.
 I. Title II. Series
 338.8'8872-dc22

Library of Congress Cataloging-in-Publication Data
Jordaan, Jacob A.
...Foreign direct investment, agglomeration and externalities : empirical evidence from Mexican manufacturing industries / by Jacob A. Jordaan.
  p. cm. -- (Ashgate economic geography series)
 Includes bibliographical references and index.
 ISBN 978-0-7546-4729-4 (hardback)
1. Business enterprises, Foreign--Mexico. 2. Investments, Foreign--Mexico. 3. Externalities (Economics) 4. Manufacturing industries--Mexico. 5. Industrial clusters--Mexico. 6. Economic development--Mexico. I. Title.

 HD2755.5.J676 2009
 332.67'30972--dc22

                                                               2009024248

ISBN 978-0-7546-4729-4 (hbk)
ISBN 978-1-138-25401-5 (pbk)

Contents

This book is dedicated to my parents

List of Figures

List of Figures

List of Tables

Preface

My interest in the operations and effects of multinational enterprises and foreign direct investment (FDI) in developing counties originates from a research trip that I took as a student in 1995 to the border region between the US and Mexico. During this trip, organised by Utrecht University, we visited a number of FDI firms in the north of Mexico. After having read many articles and positive and negative opinions on the importance of foreign-owned firms in developing countries, I found it fascinating to see first hand how these firms actually operated and what impact they appeared to generate in their local environments. In 1996, I found myself in Mexico again, working as visiting researcher at the ITESM University in Monterrey. During that year, I carried out a project on the economic impact of FDI firms in the north east of Mexico. I visited and interviewed a large number of foreign-owned firms in this region, obtaining information on and analysing a range of effects that these firms could be linked to, including direct and indirect employment effects, exporting activities, multiplier effects via the use of local suppliers of inputs and technological learning effects among Mexican firms.

In the late 1990s, I became interested in particular in the latter type of effect. I realised that if I wanted to improve my understanding of the impact of FDI in developing countries, I needed to start concentrating on more detailed aspects of this impact. One of the reasons that I started to focus on externality effects is that at the time there was a surge in quantitative research on so-called FDI spillover effects, referring to productivity effects among domestic firms in a host economy that result from the presence and operations of FDI firms. My interest in the concept of externalities was stimulated further by the growing popularity of analysing the roles of externalities in processes of growth and development at sub-national levels. In particular, I became increasingly convinced of the validity of ideas on the significance of agglomerations of economic activity as generators of place-based externalities and the importance of geographical proximity for the generation and transmission of spillover effects. As I continued to read and think about these ideas on the geography of externalities, I found it striking that the research field on FDI externalities continued to develop with only very few direct links to the literature on spatial dimensions of externality effects. I found this omission particularly striking given the important challenges that this field was (and is) facing. Despite the clear theoretical notion that FDI firms are likely to act as source of new knowledge and technologies to domestic firms in host economies, the evidence on the occurrence, nature and determinants of FDI externalities has remained very heterogeneous in nature. As a result, there is little consensus on these externalities, clearly indicating the need to develop improved research strategies.

Within this context, I started to work on the idea that the incorporation of spatial dimensions in the estimation and analysis of FDI externalities might improve the empirical identification of these effects and advance our understanding of when, why and how these effects materialise. Although several recent studies have started to explore spatial dimensions of FDI spillovers in some shape of form, they are representative of the tendency not to link into the broad literature on the geography of externalities. Perhaps the best example of this is that only a very few studies look at possible effects of industry agglomeration or geographical proximity between firms on FDI productivity effects. Empirical research on effects from agglomeration shows clearly that the tendency of firms to concentrate geographically in space can be linked to the materialisation of place-based externalities that are directly linked to the existence of an agglomeration of economic activity. Surely, such a finding hints strongly at the possibility that the externality impact of FDI firms will be affected when FDI firms operate in such an environment that is conducive to the creation and transmission of externality effects. In fact, findings from studies on FDI location behaviour contain strong indications that FDI firms are attracted to agglomerations of economic activity within host economies, suggesting that the relation between agglomeration and FDI spillovers is an important one to investigate.

The purpose of this book is to present a variety of empirical research on FDI externalities in the host economy of Mexico, whereby I place a special focus on identifying and understanding the spatial dimensions of these FDI effects. One of the questions to which I try to find an answer is whether and how agglomeration affects FDI spillovers. Also, I address the important question whether effects from FDI vary at different levels of geographical aggregation. FDI externalities are generated and transmitted via competition effects, processes of labour turnover between FDI and domestic firms, input-output linkages between FDI and domestic suppliers and via demonstration effects. As it is very likely that geographical proximity or agglomeration affects the workings of these channels to a different degree, it seems plausible to assume that estimations of FDI effects need to be carried out at different levels of geographical aggregation in order to identify better the range of externality effects from FDI. In extension of this, I also address the question whether and to what extent FDI externalities occur across geographical space. A central lesson from research on knowledge spillovers is that geographical proximity between firms enhances these spillovers. This does not mean however that inter-regional spillover effects from knowledge-creation do not occur. In fact, recent empirical evidence on the causes and effects of knowledge creation in both the EU and the US contain strong indications that such inter-regional effects do indeed materialise. Given these findings, I expect that the inclusion of inter-regional FDI effects in the estimation of FDI externalities is very likely to generate a more accurate and complete capturing of these effects.

I have been very fortunate to have received the support and advice from many colleagues and friends during the years that I have been conducting the research for this book. With the risk of forgetting people (my apologies if you are one of them!), I would like to thank the following persons in particular. First of all, I would like to thank my parents, who made it possible for me to pursue my Masters and PhD degree, during which my initial interest in FDI, geography and externalities was formed. I am also very grateful to staff members of the department of Geography and Environment at the London School of Economics and Political Science. Paul Cheshire and Gilles Duranton helped and motivated me during the design and execution stages of many parts of the research project. Together with Henry Overman, Steve Gibbons, Christian Hilber, Ian Gordon and Andrés Rodríguez-Pose, they provided many constructive and helpful comments. Special thanks go to Vassilis Monastiriotis and Eduardo Rodríguez Oreggia y Román, who provided many valuable comments over the years and became good friends during the process. I also thank Abel Pérez Zamorano, Javier Sánchez Reaza and Jorge Vera García, who were a valuable source of information on Mexico and good mates to relax with. I also appreciate the good discussion that I had with Ron Martin and Jonathan Haskel about my work and their many critical yet very constructive comments.

Next, my appreciation goes out to the people who helped me during my fieldwork in Nuevo León, Mexico. I thank the professors at the department of economics at the ITESM University in Monterrey for making me feel welcome and for their advice and support. I am also very grateful to the staff at the headquarters of INEGI in Aguascalientes for providing me with the data and helping me with the construction of reliable datasets. I also thank the students of the course in international economics that I taught at the ITESM University, who impressed me with their brightness, curiosity and dedication and who also made sure I did not feel like a tourist in my spare time.

I am indebted to Mike Taylor, who helped me tremendously during my time at Birmingham University. He stimulated me to write this book and he offered many helpful suggestions for several of the empirical chapters. I am grateful to Dominique Moran and Phil Jones, who made sure I felt at home in Birmingham. I would also like to thank the editor of *World Development* and Elsevier for permission to reuse material from two articles of mine in their journal (Jordaan 2005, Jordaan 2008c). I also thank the editor of *Environment and Planning A* and Pion Ltd. for permission to reuse material from the following article: Jordaan, J.A. (2008b) 'Regional foreign participation and externalities: New empirical evidence from Mexican regions', Vol. 40, p. 2948-2969.

My appreciation also goes out to my current colleagues at the department of economics at the VU University Amsterdam, who have been listening patiently to my stories about externalities in Mexico for two years now. Jan Willem Gunning, Chris Elbers, Remco Oostendorp and Eric Bartelsman took the time to read several parts of the empirical chapters and offered many valuable comments and suggestions.

Finally, this book could not have been written without the continued support from my partner Enikö Biró. She has, without complaints, put up with me having to work many evenings and weekends and has helped and motivated me during the entire process of writing the book, for which I am very grateful.

Jacob A. Jordaan,
Amsterdam 2009

Chapter 1
Foreign Direct Investment in Mexico: Externalities and Geographical Space

1. Introduction

The last 20 years have witnessed a revival in the interest in the central role that externalities can play in processes of economic growth and development. One reason for this increased interest is that endogenous growth theory has developed important ideas on growth effects that are linked to international flows of knowledge and technologies (Romer 1994, Grossman and Helpman 1994). As these flows of knowledge and technologies are not (fully) captured by the market, their effects occur in the form of knowledge spillovers or externalities. Furthermore, externalities as source of competitive advantage are becoming increasingly important in light of growing levels of economic integration and liberalisation in the world economy. In relation to this, ideas developed in diverse research fields including economic geography, international economics and business studies, all point at the growing regionalisation of drivers of regional growth (Rodríguez-Pose 1998, Fujita et al. 1999, Porter 1998, Rosenthal and Strange 2004). In combination, these developments have led to a marked increase in the analysis of externality-related growth processes at sub-national levels.

Although this recent upsurge in the interest in externalities is generating more attention to growth effects at the regional level, it is important to recognise that the analysis of spatial dimensions of externality effects forms part of a research tradition with an established history. It is well known that economic activity shows a persistent tendency to concentrate in geographical space (e.g. Henderson et al. 2001, Combes and Overman 2004). Explanations for this particular type of location pattern of economic activity can be traced back to the original writings by Marshall (1890), who studied the functioning of successful industrial districts within the UK. In explaining the success of these geographical concentrations of economic activity, Marshall argued that there must be benefits from agglomerating, *agglomeration economies*, which are uniquely linked to the agglomeration itself. One explanation for the occurrence of these place-based externalities is that the concentration of firms allows for the existence of thick labour markets, offering important search and match externalities to both firms and workers (Duranton and Puga 2004, Gordon and McCann 2000). Second, the agglomeration of firms promotes the existence of a pool of specialised suppliers on (non-traded) inputs (Duranton and Puga 2004), creating important proximity effects for both client and supplier firms. Also, it is likely that productivity levels are stimulated by firms becoming more inter-connected to other firms in the agglomeration, fostering increased specialisation effects (Kaldor 1970). Finally, the agglomeration of firms allows for the creation of regional experience and pools of knowledge, fostering the generation and transmission of important knowledge spillover effects (Duranton and Puga 2004, Henderson 2001, 2007). This research field has produced an impressive amount of empirical evidence on the existence and scale of these agglomeration economies, and although there are important debates on measurement and identification, the overall impression is that place based externalities can be an important source of productivity increases for firms in these agglomerations (see e.g. Eberts and MacMillen 1999, Rosenthal and Strange 2004, also Hanson 2001a).

The importance of geographical proximity in externality-generating and transmitting processes is further addressed in research on externality and growth effects from R&D and other types of innovation and knowledge creating activities (see Audretsch and Feldman 2004, Döring and Schellenbach 2006). Using the concept of the knowledge production function, early empirical studies using country and industry data were successful in identifying relationships between knowledge inputs and outputs (see Grilliches 1992, 1979). The evidence on these relationships at the micro-level proved much less convincing, however. This feature that the knowledge production function is more successfully identified at more aggregate levels strongly suggests that processes of knowledge creation involve externalities (Audretsch and Feldman 2004). Empirical research that uses regionalised versions of the production function provides robust evidence that this is the case (e.g. Jaffe 1989, also Acs, Audretsch and Feldman 1994). Innovative activity is positively influenced by regional university and corporate R&D activities, confirming both the importance of externalities as well as their regional dimensions. In extension of these findings, Jaffe et al. (1993) and Jaffe and Trajtenberg (2002) provide further corroborating evidence of the importance of geographical proximity for knowledge spillovers in the form of significant levels of localisation of knowledge creation in states in the US. Additional support for such regional dimensions of externalities is provided by research on human capital externalities which shows that regional (urban) productivity is positively affected by the regional level of human capital (see Moretti 2004).

Finally, in extension of research that focuses on the regional dimensions of externality-generating and transmitting processes, related research on innovation and knowledge creation focuses on the spatial reach of knowledge-related externalities. Although geographical proximity between economic agents is important, this is not to say that externality effects are necessarily confined at the intra-regional level. In other words, knowledge spillovers are very likely to be affected negatively by geographical distance between firms or regions, but not necessarily to the extent that inter-regional effects do not occur. Evidence on the spatial reach of knowledge spillovers is presented for the US by Anselin et al. (1997, 2000), for Germany by Bode (2004) and for EU regions by Moreno et al.

(2005), Crescenzi et al. (2007) and Rodríguez-Pose and Crescenzi (2008). These studies present evidence that clearly indicates that knowledge spillovers have an inter-regional reach. At the same time, most of the findings also indicate that these spillovers are subject to a geographical cut off point, confirming the importance of geographical proximity in these processes.

2. Foreign Direct Investment and Externalities

The research field engaged with the identification and analysis of economic effects from multinational enterprises (MNEs) and Foreign Direct Investment (FDI) constitutes another field with an increased interest in externality growth effects. FDI-induced externalities or spillovers refer to situations where the level of efficiency or productivity of domestic firms in a host economy is affected by the presence and operations of FDI firms, effects for which the foreign-owned firms are not (fully) compensated (Blomström and Kokko 2003, 1998, Caves 1996, Venables and Barba Navaretti 2005). It is this feature of lack of compensation that makes these FDI effects externalities. One explanation for the occurrence of these effects is that the presence of FDI firms increases the level of competition on the host economy market, forcing domestic firms to become more efficient. Another explanation is related to processes of inter-firm labour turnover. When a worker substitutes a domestic firm for a foreign-owned firm, the domestic firm can benefit from the skills and experience that the worker gained while working for the foreign-owned firm. Input-output linkages between FDI and domestic supplier firms may also transmit externalities, especially in cases where FDI firms provide support to these suppliers. Finally, the presence of FDI firms may generate demonstration effects, where domestic firms learn about new technologies from observing foreign owned firms and their products and through informal face-to-face contacts, regional business associations and trade journals.

Although there is a clear consensus on the theoretical notion that FDI firms can generate important positive externalities in host economies, the literature shows that this research field is facing several challenges in obtaining corroborating empirical evidence on these externalities. One important debate centres on the question how important positive FDI spillovers actually are. Cross-country growth regressions provide little evidence in support of a general positive FDI growth effect (Caves 1996, Lipsey 2004, Carkovic and Levine 2005). As for applied micro-economics research, whereas some argue that there is substantial evidence of positive intra-industry FDI externalities (e.g. Blomström and Kokko 2003, 1998, Ewe-Ghee Lim 2001), others disagree strongly with such a positive interpretation of the evidence (e.g. Hanson 2001b, Rodrik 1999, Aitken and Harrison 1999). Findings that suggest that FDI firms may also generate negative intra-industry externalities (Aitken and Harrison 1999, Konings 2000) further indicate the complexity of the issue.

In response to this impasse, recent research has started to explore new avenues to obtain improved empirical evidence of these FDI effects. One development is that there is a rapid increase in the number of host economies for which FDI externalities are estimated; this has not decreased the large degree of heterogeneity of the evidence, however. More importantly, a number of studies try to capture the full industry dimension of FDI externalities, by distinguishing between intra- and inter-industry FDI spillovers. Although several of these studies present evidence of positive FDI externalities of an inter-industry nature, again the evidence appears to be too diverse to conclude on the generality of this type of effect. Finally, in line with findings from cross-country growth studies that FDI growth effects may be contingent on host economy conditions (e.g. Balasubramanyam et al. 1996, Borensztein et al. 1998), micro-economics oriented research is also attempting to identify structural factors that influence the materialisation of FDI spillovers (see Blomström and Kokko 2003, Crespo et al. 2007). These factors include FDI characteristics such as nationality (Haskel et al. 2007), the level of foreign participation in FDI firms (Blomström and Sjöholm 1999, Javorcik and Spatareanu 2003) and the motivation of FDI firms to locate in a host economy (Driffield and Love 2007). The findings from these studies do suggest that such factors may play an important role, but the amount of empirical evidence has remained limited thus far.

The one factor that is more commonly accepted as being important for FDI externalities to occur is the level of absorptive capacity of domestic firms. This concept points at the requirement that domestic firms need to possess a sufficient level of technological development in order to benefit from externality effects. For instance, demonstration effects will only materialise when domestic firms possess sufficient knowledge and skills to understand and learn from new technologies that are incorporated in FDI firms. Despite of this idea being straightforward, empirical research is hampered by the problem that researchers have to use indirect indicators of the level of absorptive capacity. The most commonly used indicator is the level of technological differences between FDI and domestic firms, which is interpreted as a direct inverse indicator of the level of absorptive capacity of domestic firms. The problem with this is two-fold. First, it is not clear whether this interpretation of the technology gap is based on a complete interpretation of the underlying original catch up thesis, which was developed to understand the role of technology spillovers between leading and lagging countries as stimulator of growth of the lagging countries (Jordaan 2005, 2008c). Second, several studies present evidence that suggests that large technological differences between FDI and domestic firms are conducive rather than detrimental to FDI externalities, which is in direct contrast to the traditional interpretation of what the technology gap captures. Studies that find such an opposite effect of the technology gap do not provide a satisfactory explanation for this effect, however, reinforcing the impression that there are problems with the empirical translation of the concept of absorptive capacity into the level of technological differences between FDI and domestic firms.

Spatial Dimensions of FDI Externalities

It seems clear that the research field on FDI externalities is facing several important challenges. As such, there is a strong need to continue to pursue new research strategies to obtain improved evidence on the occurrence and nature of these effects and to develop a better understanding of the factors that may be conducive or detrimental to the occurrence of these externalities. It is also clear that research on relations between geographical space and externalities shows that spatial dimensions can play important roles in processes that generate and transmit spillover effects. In combination, this suggests that contemporary research on FDI effects may stand to gain substantially from incorporating and analysing spatial dimensions of FDI externalities. An important indication that relations between FDI effects and agglomeration or geographical proximity between firms need to be addressed can be found in research on FDI location decisions. Several studies provide important evidence that FDI firms are attracted to regions within host economies that contain agglomerations of economic activity (e.g. Head et al. 1995, 1999, Crozet et al. 2004, Hilber and Voicu 2009). Not only do such findings indicate that agglomerative forces appear to be an important element in the business environment of FDI firms, they also suggest that the relation between agglomeration and FDI spillovers is an important one to analysis, given the apparent preference of FDI firms for operating in agglomerations of economic activity in host economies.

There are several aspects of the relation between geographical space and FDI externalities that can be explored. First, concerning the possible relation between agglomeration and FDI effects, it is important to recognise that there is a substantial degree of similarity between the channels that underlie FDI spillovers and the mechanisms that generate agglomeration economies (Jordaan 2008c). As both FDI externalities and agglomeration economies are related to processes of labour turnover, input-output linkages and knowledge spillovers, the hypothesis that FDI effects are facilitated by agglomeration or geographical proximity is easily proposed. Channels of FDI externalities are more likely to come into existence and be more effective when FDI firms operate in an agglomeration of economic activity. Also, the high level of geographical proximity between FDI and domestic firms that follows from co-location in an agglomeration will have a general enhancing effect on the generation and transmission of knowledge spillovers. Second, the literature on the regionalisation of externality effects from R&D and innovations suggests that there may also be an important degree of regionalisation of FDI effects. Translating this recognition of the regional dimension of FDI effects into empirical estimation strategies, estimations are likely to benefit from distinguishing between national level and regional FDI effects within a host economy. In fact, given the variety of channels that may cause and transmit FDI externalities, it is very likely that there will be variation in the geographical scale at which FDI productivity effects occur. Finally, just as knowledge spillovers from R&D and innovations may not necessarily be confined at the intra-regional level, externalities from FDI in a given region may similarly affect domestic firms in other regions of a host economy. This suggests that improved identification strategies of FDI externalities are likely to require the incorporation of such inter-regional effects, ensuring that the full range of FDI effects is captured.

FDI Externalities in Mexico

In this book, I develop a variety of empirical estimations and analysis of FDI externalities in the manufacturing sector of the republic of Mexico. Although there is substantial evidence of FDI externalities in this host economy, the relevance of this evidence for understanding the role of FDI in contemporary processes of economic growth in this host economy appears to be constrained. The central limitation is that evidence on FDI productivity effects is based on the use of data from the 1970s and 1980s (e.g. Blomström en Persson 1983, Blomström 1989, Kokko 1994, Blomström et al. 2000, Aitken et al. 1997, Grether 1999). Although these studies have made significant contributions to our understanding of FDI externalities in Mexico, the 1970s and most of the 1980s were characterised by the presence of stringent policies of import substitution and government intervention. As the Mexican government introduced a new development strategy based on economic liberalisation and trade promotion in the late 1980s, generating a dramatic influx of new FDI into its economy, there is a clear need to obtain evidence on FDI externalities in this host economy for more recent periods. Furthermore, the Mexican case offers a particularly interesting setting to analyse FDI externalities, given the substantial spatial changes that economic activity in Mexico experienced following the introduction of trade liberalisation (see Hanson 1998a, Jordaan and Sánchez-Reaza 2006, Faber 2007). In particular, the geographical distribution of economic activity changed from consisting of one main agglomeration of economic activity in Mexico City to one where a limited number of regional production centres in the north and the centre of the country now incorporate important shares of the Mexican economy. The growing importance of these agglomerations of economic activity underlines the importance of analysing relations between agglomeration and FDI externalities in this host economy.

The purpose of this book is to conduct a variety of empirical analysis of FDI externalities in the Mexican economy, to obtain new evidence on the effect of the technology gap and to explore the spatial dimensions of these productivity effects. I built several datasets from both published and unpublished data, provided by the Instituto Nacional de Estadística y Geografía (INEGI), which is the main government institution responsible for all the main censuses and surveys. Furthermore, I also spent a considerably long time in the state Nuevo León, located in the north east of Mexico and representing the second largest agglomeration of economic activity after Mexico City. In this region, I carried out large scale firm level surveys among both Mexican and foreign-owned producer firms and local suppliers. Although the process of data gathering and database building proved to be very labour intensive, costly and time consuming, I believe that the datasets allow me to obtain important new empirical evidence on the operations and externality effects of FDI firms. For instance, despite the very important role that FDI firms play in the Mexican economy, very little is known about the factors that influence the regional distribution of FDI firms within Mexico. Using thus far unexplored data, I conduct a detailed quantitative analysis of FDI location decisions during the 1990s, paying special attention to the role of agglomeration economies as location factor. Also, the research approach that I develop is more wide-reaching compared to research on many other host economies. In particular, I estimate FDI externalities at both the intra- and inter-industry level, at different levels of industry and geographical aggregation. I estimate regional FDI spillover effects within the context of regional growth during the last two decades, FDI externalities at the national and regional level using detailed industry data and FDI effects at both the intra- and inter-regional level. In all these estimations, I systematically explore the spatial dimensions of these FDI effects, as well as the role and importance of the technology gap. Also, I have taken great care to obtain unbiased and representative firm level information on the level and nature of local sourcing patterns by producer firms in Nuevo León. To the best of my knowledge, the case study on Nuevo León is the first where the findings are representative for both domestic and foreign-owned producer firms, as well as for the local supplier base. As a result of this variety of approaches, the empirical chapters present a wide range of new evidence on the existence and nature of FDI externalities and offer important insights that will contribute to our understanding of these FDI effects and their spatial dimensions. As such, the findings will not only improve our understanding of FDI effects in the host economy Mexico, they will also make an important contribution to the wider literature on the operations and effects of FDI firms.

3. Structure of the Book

Following this introductory chapter, the book consists of six chapters and a concluding chapter. In Chapter 2 I introduce the concept of FDI externalities and discuss more in depth the channels through which these effects can materialise. Furthermore, I survey the body of existing evidence on FDI externalities, focusing on findings of intra- versus inter-industry effects, evidence on the importance of absorptive capacity of domestic firms as indicated by findings on the effect of the technology gap and findings from a number of recent studies that have started to incorporate some form of assessment of the spatial dimensions of FDI externalities.

Chapter 3 consists of two parts. In the first part I discuss the main characteristics of inward FDI in the Mexican economy. I focus on the period of economic liberalisation and trade promotion, as this period is characterised by a dramatic increase in the level of inward FDI. I discuss the allocation of FDI over the various economic sectors and the regional distribution of FDI firms. Of course, this section also discusses the developments of the maquiladora industries, as the large presence of maquiladora firms forms a central feature of the level and nature of foreign participation in the Mexican economy. The second part of this chapter presents new empirical findings on determinants of the regional distribution of new FDI firms, based on thus far unexplored FDI location data for the 1990s. The analysis is focused in particular on determining whether and how agglomerations of economic activity within Mexico influence the location behaviour of new FDI firms. Evidence that these firms are attracted to regions that contain agglomerations of economic activity would clearly underline the importance of analysing relations between agglomeration and FDI effects in Mexico.

Chapter 4 presents an empirical analysis of regional FDI growth effects, within the context of the marked locational changes that economic activity in Mexico has undergone following the introduction of trade liberalisation in the 1980s. In the first part of this chapter, I describe these locational changes in some detail and discuss empirical evidence on the nature, scale and determinants of these changes. Importantly, previous research has paid only limited attention to the potentially important role of FDI as central factor in these processes of spatial adjustment, which is the more remarkable given the large increase in the level of inward FDI in Mexico during the years of trade liberalisation. In response to this gap in the literature, the second part of this chapter presents a new empirical study on determinants of regional growth, which is focused in particular on identifying growth effects from agglomeration economies and regional foreign participation. In extension of this, the analysis also looks at the magnitude and nature of inter-regional growth effects from agglomeration and FDI.

Chapter 5 focuses on the empirical identification of intra-industry FDI externalities in Mexican manufacturing industries. For the empirical analysis in this chapter I use two datasets that I built with unpublished data from the 1994 manufacturing census. One dataset contains observations for 6-digit manufacturing industries at the national level, whereas the other dataset contains similar information for a number of Mexican states that have a relative large share in the Mexican economy. Given the cross-sectional nature of these datasets, an important section of this chapter discusses the problem that cross-sectional estimations of FDI spillovers may be affected by problems of causality or endogeneity of the variable that measures intra-industry foreign participation. To deal with this problem, I introduce a new instrument for this problematic right hand side variable which allows me to conduct unbiased estimations of intra-industry FDI externalities. I address the regional dimension of these FDI externalities by estimating for FDI effects at both the national and the regional level. In extension of this, I use both datasets to obtain new indications of the extent and nature of the effects of the technology gap and industry agglomeration on FDI externalities.

Chapter 6 extends the scope of the empirical identification strategy by estimating for both intra- and inter-industry externalities and by exploring more fully the spatial dimensions of these FDI effects. For the empirical analysis, I continue to use the dataset with selected Mexican states and I use an additional dataset containing more aggregated industry observations for all 32 Mexican states. With these datasets, I investigate whether the technology gap and industry agglomeration affect intra- and inter-industry FDI externalities, both at the national and the regional level. Furthermore, I conduct estimations to identify inter-regional FDI effects and I assess whether there are regional characteristics that may influence these spatial externalities.

In Chapter 7, I analyse unique firm level data that I obtained from surveys among producer firms and local suppliers in the state Nuevo León. These surveys are focused in particular on obtaining detailed information on the extent to which producer firms use local suppliers of inputs and the extent and nature of support that producer firms provide these local suppliers with, as such support can be linked to the occurrence of positive externalities. In the empirical analysis, I take great care to compare Mexican and foreign-owned producer firms, to assess whether FDI firms are actually different from domestic firms in the level and nature of their local linkages. Importantly, the surveys allow me to assess the extent and importance of local linkages and support from the point of view of both supplying and receiving parties. In extension of the dichotomous comparisons between FDI and Mexican producer firms, I also conduct statistical analysis to identify the effects of a variety of firm level characteristics on both the level and nature of local linkages that producer firms have established with local suppliers.

Finally, in Chapter 8 I summarise the main findings, discuss policy implications and speculate on some future research directions.

Chapter 2

FDI Externalities in Host Economies: Heterogeneity, Technology Gap and Geographical Space

1. Introduction

MNEs play a central role in the contemporary world economy, by engaging in growing levels of FDI in many host economies (Venables and Barba Navaretti 2005). Furthermore, not only have most countries abolished or at least severely relaxed policies restricting inward FDI into their economies, the promotion and active attraction of new foreign-owned firms often constitutes an important element of economic development policies in many developed and developing countries. The reason for the popularity of attracting international investment is that FDI firms can generate a variety of positive economic effects, including capital investment growth effects, direct and indirect employment creation, the generation of new revenues from international trade, multiplier effects caused by inter-firm linkages between FDI firms and supplier and client firms and technology enhancements through formal technology transfers between foreign-owned and domestic firms.

In the last two decades, there has been a growing recognition of another element of the positive economic impact that FDI firms can generate. This effect occurs in the form of technological externalities, which arise when the efficiency or productivity of domestic firms is affected by the entrance and operations of foreign-owned firms, effects for which FDI firms are not (fully) compensated. Although this type of effect has been the subject of research for a considerable time, it is especially in the last 20 years that there has been an impressive growth in quantitative research that tries to identify and quantify these spillover effects. In particular, following initial macro-level studies that attempted to identify positive FDI growth effects from aggregate cross-country growth regressions, improvements in the availability and quality of datasets has fostered the development of more micro-economics oriented research on FDI externalities in individual host economies.

The purpose of this chapter is to provide a fresh survey of the substantial body of empirical research on FDI externality effects. The reason for conducting this new survey is two-fold. First, although several surveys have been conducted in recent years (e.g. Blomström and Kokko 1998, 2003, Görg and Strobl 2001, Görg and Greenaway 2004, Lipsey and Sjöholm 2004), there appear to be substantial differences in the interpretation of the evidence. Whereas some argue that there is sufficient evidence to support the notion that positive FDI externalities are an important phenomenon (Blomström and Kokko 2003, 1998, Ewe-Ghee Lim 2001), others disagree strongly and argue that the evidence is in fact rather weak (Hanson 2001b, Rodrik 1999). Evidence that suggests that the presence and operations of FDI firms may also generate negative externalities among domestic firms in host economies further indicates the complexity of the issue (see e.g. Aitken and Harrison 1999, Konings 2000).

Second, the research field on FDI spillovers is characterised by important recent challenges and developments which have not received sufficient attention and scrutiny thus far. For instance, several studies have started to distinguish between intra- and inter-industry FDI spillovers. Although this is recognised in previous surveys, most of the recent studies that make this distinction are not included in these surveys. Also, recent research is trying to obtain new evidence on the importance of endogenous factors that may affect the materialisation and perhaps also the nature of FDI spillover effects. The best known example of this is research that estimates the effect of the level of technological differences between FDI and domestic firms on FDI externalities, whereby this level of technological differences is interpreted as a direct inverse indicator of the level of absorptive capacity of domestic firms. Although the importance of a sufficient level of absorptive capacity for the materialisation of spillovers is undisputed, the use of the technology gap between FDI and domestic firms as indicator of the capacity of the latter type of firm to absorb new knowledge and technologies has remained largely un-scrutinised in the literature thus far. Furthermore, in the context of developing new strategies to both improve the empirical identification of the full range of FDI effects and to identify new determinants of these effects, several studies have started to look at spatial dimensions of FDI spillovers. Although clearly important, these studies have received only limited attention in previous surveys.

To conduct the new survey, the chapter is constructed as follows. In section 2, I discuss the concept of externalities and explain the various mechanisms that may link the presence and operations of foreign-owned firms in host economies to the occurrence of externality effects. Section 3 provides a survey of the overall findings from a large set of micro-economics empirical studies on FDI effects. In section 4, I discuss the use of the technology gap as inverse indicator of the level of absorptive capacity of domestic firms and look at the available evidence. Section 5 presents the main arguments to include spatial dimensions in applied research of FDI spillovers and reviews the available evidence on these dimensions. Section 6 summarises and concludes.

2. FDI-induced Externalities

Externalities can be thought of as consisting of two types, namely technological externalities and pecuniary externalities (Viner 1953, Scitovsky 1954, Mishan 1971, Papandreou 1994). Technological externalities occur when '...the actions of one agent directly affect the environment of another agent, i.e. the effect is not transmitted through prices' (Papandreou 1994: 5, see also Bator 1958). Suppose there are two firms A and B. Firm A introduces a new technology in its production process. If firm B learns about this new technology and uses this knowledge to improve its own production process, a positive technological externality has occurred, provided that firm B does not have to compensate firm A for the improvement of its production process. This lack of compensation is caused by the absence of a market mechanism that captures the flow of knowledge from firm A to firm B.

In contrast to this process, pecuniary externalities are effects that are transmitted through the market. They are said to arise '...whenever the profits of one producer are affected by the actions of other producers' (Scitovsky 1954: 146, see also Ottaviano and Puga 1998). Suppose again that there are two firms A and B and that firm B uses the product of firm A as input in its production process. Also assume that the production of firm A is subject to increasing returns to scale. This means that a change in the production volume of firm A will alter its efficiency level. This will change the price that firm B has to pay for the input that is produced by firm A, thereby affecting the profit level of firm B. In other words, the actions of firm A (i.e. change in production volume) can effect the efficiency level of firm B; however, in this case the market mechanism does pick up the effect.

Empirical research on externalities from FDI appears to focus on technological externalities. In broad terms, these effects can be described as 'all those phenomena tied to the presence of foreign firms on the national territory that may increase the productive efficiency of domestic firms' (Perez 1998: 22). The reason for interpreting these phenomena as technological externalities is that they are unintended and that the foreign-owned firms are not compensated for their creation (Perez 1998). Related definitions are proposed by Caves (1974, 1996) and Blomström and Kokko (1998, 2003). Caves states that FDI externalities occur 'when...[the] multinational corporation can not capture all the quasi rents due to its productive activities, or to the removal of distortions by the subsidiaries' competitive pressure' (Caves 1974: 176). Along similar lines, Blomström and Kokko (1998) state that spillovers from FDI occur '...when the entry or presence of MNE affiliates leads to productivity or efficiency benefits in the host country's local firms, and the MNEs are not able to internalise the full value of these benefits' (Blomström and Kokko 1998: 3).

These definitions appear to refer exclusively to cases of technological externalities, as they stress the lack of compensation for effects that FDI firms may generate, caused by the absence of a market mechanism that captures these effects. However, the definitions do not necessarily exclude cases of pecuniary externalities. For instance, Caves (1974) refers to the change in competitive pressure in a host economy that follows from the entrance of FDI. This effect will be 'normally' transmitted through the market via price signals, for instance when a foreign-owned firm offers a product at a price below the price charged by domestic firms. Domestic firms may respond to this price signal by improving their efficiency levels, in order to meet the new lower price. Another example is when a FDI firm establishes input-output linkages with domestic suppliers, where the foreign-owned firm exerts pressure on these suppliers to accept lower prices for their products. If the suppliers respond to this pressure by improving their production process, the presence of the FDI firm results in an enhanced efficiency level among these domestic firms. As these effects are transmitted through the market, the resulting changes in efficiency levels of domestic firms resemble cases of pecuniary rather than technological externalities. Most empirical studies on FDI externalities adopt explicitly or implicitly definitions such as those proposed by Caves (1974) and Blomström and Kokko (1998). The main consequence of this is that, at least until recently, the focus of empirical research has rested solely on the identification of externality effects from FDI in the form of positive technological externalities. However, it appears that FDI can also be linked to externality effects of a pecuniary nature. In essence, technological externalities arise when foreign-owned firms act as source of new knowledge and technologies, whereas pecuniary externalities arise when the presence and operations of FDI firms lead to a change in conduct among domestic firms in a host economy.

Broadly speaking, the operations of MNEs and FDI can affect the level of technology in a host economy in two main ways. The first is through formal technology transfers between a MNE or FDI and domestic firms. Second, the presence and operations of FDI may generate FDI-induced externalities. Figure 2.1 depicts these two main sources of technology, as well as the variety of ways through which externalities can be transmitted from FDI to domestic firms.

MNEs can disseminate technologies at the international level via formal technology transfers in two main ways.[1] One type of formal technology transfer is the internal flow of knowledge and technologies that occurs within the structure of a MNE from the home country to its affiliates in host economies. As an indication of the importance of this type of technology transfer, estimations for the 1980s and 1990s indicate that between 67 and 90 percent of international technology flows were intra-firm transfers (Saggi 2002). The other form of formal technology transfer is external and concerns the sale of technologies by either a MNE or its affiliates to domestic firms in a host economy. As a result of both these types of formal technology transfer, the overall level of technology in the host economy will increase.

---

1   The central role of MNEs in the creation of new technologies is indicated by the fact that MNEs account for about 65 percent of total expenditure on business R&D (see UNCTAD, 2005, 1992; Blomström et al., 1999).

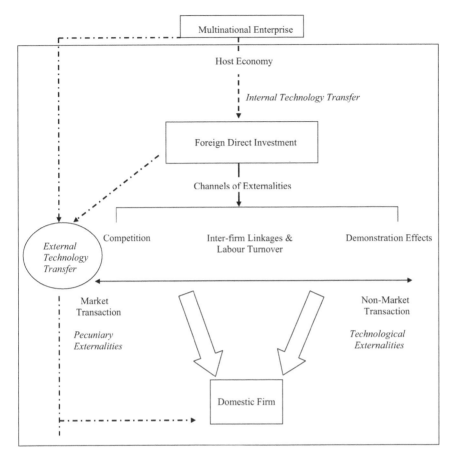

Figure 2.1   Technology transfers and FDI externalities

Next to these formal technology transfers, the presence and operations of FDI firms can generate externality or spillover effects. These externalities can be transmitted via several channels, which are depicted in Figure 2.1 alongside an axis indicating the extent to which the externalities that are transmitted consist of pecuniary or technological externalities. On the far left of the axis are placed formal technology transfers, as these are technology flows that are fully captured by the market. Next to these formal technology transfers are depicted FDI externality effects that are caused by the competition effect. The traditional notion is that the entrance of FDI enhances the level of competition in a host economy, thereby stimulating domestic firms to become more efficient. According to this interpretation, the competition effect results in the creation of positive pecuniary FDI externalities. Having said so, recent findings that indicate the materialisation of negative FDI externalities point at the possibility that the competition effect may also be negative. These negative FDI externalities are explained by the occurrence of a so-called market stealing effect (Aitken and Harrison 1999, see also Venables and Barba Navaretti 2005). The entrance of FDI firms lowers the market share of domestic firms. If the production of these firms is subject to scale economies, the decrease in production volume that results from the smaller market share will reduce efficiency, representing a case of negative pecuniary FDI externalities.

In the centre of the axis that indicates the nature of FDI externalities are depicted the externality-transmitting channels of inter-firm linkages and labour turnover. Both these channels represent mechanisms that may transmit FDI externalities of a technological and/or pecuniary nature. Looking first at the channel of inter-firm linkages, pecuniary externalities can arise when the production of a local supplier of a FDI firm is subject to increasing returns to scale. When the FDI firm increases its demand for the product of the local supplier, the resulting increase in production volume of the supplier will lower the price of its product. Furthermore, it is not only the FDI firm that will benefit from this price decrease, but also other firms that purchase the product from the supplier. Another way how input-output linkages can transmit pecuniary externality effects is when a FDI firm exerts pressure on a local supplier, for example by demanding lower prices or better quality products. Such pressure can force the supplier to change its conduct and increase its efficiency, representing a case of positive pecuniary externalities.

Technological externalities can be transmitted through inter-firm linkages when FDI firms act as a source of new technologies and knowledge to their local suppliers. It is important to recognise that the majority of input markets do not resemble markets of perfect competition (Lall 1980, Dunning 1993, see also UNCTAD 2001). For instance, there may be a limited number of buyers and/or sellers and products may be very heterogeneous and subject to frequent change. Under such conditions, inter-firm linkages involve extensive contacts between buying and selling parties, fostering the frequent exchange of knowledge and information. Furthermore, especially in the case of FDI in developing countries, FDI firms may be actively involved in improving their local suppliers, trying to obtain better quality inputs, price decreases or improvements in the reliability and speed of delivery (UNCTAD 2001). To achieve this, FDI firms may offer a variety of types of support to local suppliers, such as assistance with quality control processes, the provision of special machinery and tools and assistance with the introduction of new production technologies (Lall 1980, Dunning 1993, UNCTAD 2001, Potter et al. 2002, Javorcik 2008). When the support offered by a foreign-owned firm outweighs the concessions that it receives in return from the suppliers (e.g. in the form of lower prices), these supportive linkages result in the occurrence of positive externalities. As it may be very difficult in practice for FDI firms to obtain complete compensation for their support, the common assumption is that these supportive linkages are very likely to generate positive externality effects of some magnitude (Blomström and Kokko 1998, UNCTAD 2001).

The other channel that can transmit both pecuniary and technological externalities is the channel of labour turnover or inter-firm labour mobility.

Technological externalities arise when workers substitute a domestic for a foreign owned firm, enabling the domestic firm to benefit from the skills and knowledge that these workers gained while working for the foreign-owned firm. As for pecuniary externalities, several studies have shown that FDI firms pay higher wages compared to their domestic counterparts (Lipsey and Sjöholm 2004, Caves 1996). This wage difference usually holds even after controlling for differences in skills, productivity and profitability. This suggests that foreign-owned firms pay a wage premium to their workers, designed to lower the willingness of these employees to move to other firms (Fosfuri et al. 2001). This wage premium is related to pecuniary externalities in two ways. First, it transforms potential technological externalities into pecuniary externalities. The additional skills incorporated in employees of FDI firms are available to domestic firms, at a price equalling the wage premium. Second, the presence of FDI firms that pay higher wages puts upward pressure on industry wage levels, which will have a negative effect on domestic firms' profits. This may spur these firms to become more efficient, resulting in the materialisation of positive pecuniary externalities.

Finally, demonstration effects are placed on the far right of the axis of market versus non-market transactions, indicating that these effects are pure technological externalities. Domestic firms may learn about new technologies and production techniques when FDI firms enter a host economy. Also, learning effects and knowledge spillovers may arise through informal face-to-face contacts, meetings through business organisations and the dissemination of information through trade journals. Another type of demonstration effect leads to so-called market access spillovers (Blomström and Kokko 1998, Görg and Greenaway 2004). These spillovers occur when domestic firms increase their international trade activities as a result of knowledge and experience that they obtain from FDI firms on how to operate on international markets.[2]

3. Quantitative Evidence of FDI Externalities

Quantitative research on FDI effects consists of macro-oriented and micro-oriented approaches. Macro-oriented research on FDI growth effects was reinvigorated following the development of endogenous growth theory (Romer 1994, Grossman and Helpman 1994). In contrast to the limited long term growth effect of FDI in neo-classical growth theories, FDI is given a more important and structural role when growth determinants are taken to be endogenous (de Mello 1997). As FDI can be interpreted to consist of a bundle of capital stocks, special skills and technologies (Balasubramanyam et al. 1996, Caves 1996), the entrance of foreign-

---

2   In this chapter, I focus on FDI externalities that materialise as productivity effects. See Görg and Greenaway (2004) for a discussion of the findings of studies on FDI-induced market access spillovers.

owned firms may serve as a central source of new knowledge and technologies for host economies, generating important growth and productivity effects.

Despite this clear theoretical notion that FDI firms can generate important externality effects, the macro-economic evidence of these FDI effects is far less convincing. Time series evidence on the long run relation between inward FDI and economic growth is very heterogeneous of nature. For instance, Zhang (2001, 1999) presents time series evidence on the long run relation between inward FDI and GDP growth for 11 Latin American and Asian countries. The estimations identify a long run positive relation between these two variables for only five of these countries. De Mello (1999) looks at the long run relation between FDI and GDP growth for a sample of OECD member and non-member countries and finds that a significant positive relation between FDI and GDP growth only materialises in the latter group of countries. In contrast, Hansen and Rand (2006) present evidence of a strong causal relation between FDI and GDP for a sample of developing countries.[3]

Estimations of FDI growth effects in cross-sectional or panel data settings provide very little evidence of a general positive growth effect from FDI. Positive growth effects may occur, but only when host economies meet certain requirements. For instance, Balasubramanyam et al. (1996) finds that a positive FDI growth effect only emerges among those developing countries in their sample that are sufficiently open to international trade. Blomström et al. (1994) conduct a similar study and find that only those host economies that have reached a sufficient level of overall development appear to enjoy positive FDI effects. Findings presented by Borensztein et al. (1998) suggest that the materialisation of positive FDI effects depends on the level of human capital in host economies. Durham (2004), Hermes and Lensink (2003) and Alfaro et al. (2004) find that FDI growth effects only appear to materialise in host economies with sufficiently developed institutions and financial markets.

The evidence from these macro-economic studies does not support the theoretical notion that international investment is likely to have a universal positive long term growth effect. Furthermore, this type of research is hampered by several important problems. One issue is that the nature of the data makes it very likely that the estimations are affected by various types of aggregation bias and measurement problems. Furthermore, cross-country growth regressions in particular can be influenced by omitted variable bias (see Levine and Renelt 1992, also Sala-i-Martin 1997). Finally, there is the structural problem that it is not possible to determine *a priori* a uni-directional line of causation running from FDI to GDP or GDP growth. In fact, several time series studies identify a line of causation running from GDP to FDI, indicating that FDI gravitates towards countries with high levels of income or economic growth (Nair-Reichert

---

3   Additional time series evidence that reflects the heterogeneous nature of the relation between GDP and FDI is presented by Kholdy (1995), Chowdury and Mayrotas (2006), Cuadros et al. (2004) and Basu et al. (2003).

and Weinhold 2001). Recent findings presented by Carkovic and Levine (2005) are very important in this respect. They estimate FDI growth effects for several samples of countries, employing a variety of econometric specifications and taking great care in controlling for problems of causality and other statistical issues (see Carkovic and Levine 2005). In all of their estimations, the estimated effect of FDI remains insignificant, indicating the complexity of the empirical identification of FDI growth effects in macro-economic settings.

Micro-economics Studies of FDI Externalities

The majority of empirical research on FDI externalities consists of industry or plant level studies for individual or small groups of host economies. Especially in the last 15 years, due to the growing availability of new datasets, there has been a large increase in the number of these micro-economics studies. Table 2.1 presents the main findings and characteristics of a large majority of such studies on FDI spillover effects.

The original contribution to this type of research is represented by Caves (1974), who tried to assess whether the presence of FDI in Australian manufacturing industries was generating intra-industry externality effects. Using a cross-section of industries, Caves estimated a regression model on determinants of value added per worker in industry shares owned by Australian firms, including the share of FDI in the number of industry employees as one of the right hand side variables. The estimations produced a significant positive association between the dependent variable and this intra-industry foreign participation variable, which Caves interpreted as evidence of the existence of positive FDI externalities. This estimation strategy is characteristic for micro-economics research on FDI spillovers. As Blomström and Persson (1983) explain, '...if there is a positive relation between the productivity level of the domestically-owned plants in an industry and the share of foreign participation in the same industry (ceteris paribus), the foreign investment does raise the productivity in domestically owned plants through spillover efficiency' (Blomström and Persson 1983: 495).

A first wave of empirical studies on FDI spillover effects followed the initial contribution by Caves (1974), all using industry or plant level cross-sectional data. The findings from these studies are presented in the top half of Table 2.1. The majority of these studies present evidence that indicates the presence of positive intra-industry externalities. In particular, several studies on Mexico and Indonesia have estimated a variety of empirical models on determinants of productivity of domestic firms, all finding positive productivity effects of the level of industry foreign participation (Blomström and Persson 1983, Blomström 1986, Blomström and Wolff 1994, Kokko 1994, Sjöholm 1999, Blomström and Sjöholm 1999). Evidence of a negative association between labour productivity and industry FDI in a cross-sectional framework is presented by Haddad and Harrison (1993) for Morocco and Zukowska-Gagelmann (2000) for Poland.

Table 2.1   Main empirical findings on FDI externalities

Authors	Host economy	Type	Data	Year	Dependent Variable	Intra-Industry Externalities	Backward Linkages
Cross Sectional Studies							
Caves (1974)	Australia	Developed	Industry	1966	VA/L	(+)	–
Globerman (1979)	Canada	Developed	Industry	1972	VA/L	(+)	–
Blomström and Persson (1983)	Mexico	Developing	Industry	1970	VA/L	(+)	–
Blomström (1986)	Mexico	Developing	Industry	1970	Frontier	(+)	–
Haddad and Harrison (1993)	Morocco	Developing	Industry	1985-89	VA/L	(-)	–
Blomström and Wolff (1994)	Mexico	Developing	Industry	1970-75	VA/L	(+)	–
Kokko (1994)	Mexico	Developing	Industry	1970	VA/L	(+)	–
Kokko (1996)	Mexico	Developing	Industry	1970	VA/L	(+)	–
Kokko et al. (1996)	Uruguay	Developing	Plant level	1988	VA/L	(?)	–
Imbriani and Reganati (1997)	Italy	Developed	Industry	1988	VA/L	(+)	–
Sjöholm (1999)	Indonesia	Developing	Plant level	1980-91	VA/L	(+)	(+)
Blomström and Sjöholm (1999)	Indonesia	Developing	Plant level	1991	VA/L	(+)	–
Chuang and Lin (1999)	Taiwan	Developing	Plant level	1991	VA/L	(+)	–
Zukowska-Gagelmann (2000)	Poland	Transition	Industry	1993-97	TFP	(-)	–
Dimelis and Louri (2002)	Greece	Developed	Plant level	1997	VA/L	(+)	–
Kokko et al. (2001)	Uruguay	Developing	Plant level	1988	VA/L	(+)	–
Liu (2001)	China	Emerging	Industry	1995	VA/L	(+)	–
Schoors en van der Tol (2002)	Hungary	Transition	Plant level	1997-98	VA/L	(+)	–
Kathuria (2002)	India	Emerging	Plant level	1989-90, 1996-97	Frontier	(-)	–
Svejnar et al. (2007)	17 CEE countries	Transition	Plant level	2002-05	Prod function	(+)	(+)
Panel Data Studies							
Haddad and Harrison (1993)	Morocco	Developing	Plant level	1985-89	Frontier	(+)	–
Haddad and Harrison (1993)	Morocco	Developing	Plant level	1985-89	Prod function	(?), (-),	–
Grether (1999)	Mexico	Developing	Plant level	1984-88	Frontier	(-)	–

Table 2.1 continued   Main empirical findings on FDI externalities

Panel Data Studies							
Authors	**Host economy**	**Type**	**Data**	**Year**	**Dependent Variable**	**Intra-Industry Externalities**	**Backward Linkages**
Aitken and Harrison (1999)	Venezuela	Developing	Plant level	1976-89	Prod function	(-)	–
Konings (2000)	BL, RO, PO	Transition	Plant level	1993-97	Prod function	BL, RO(-) PO (?)	–
Djankov and Hoekman (2000)	Czech Republic	Transition	Plant level	1992-96	Prod function	(-)	–
Barrios (2000)	Spain	Developed	Plant level	199-94	Prod function	(?)	–
Kinoshita (2001)	Czech Republic	Transition	Plant level	1995-99	TFP	(?)	–
Driffield (2001)	UK	Developed	Industry	1989-92	Prod function	(?)	–
Sgard (2001)	Hungary	Transition	Plant level	1992-99	Prod function	(+)	–
Girma et al. (2001)	UK	Developed	Plant level	1991-96	Prod function	(?)	–
Kathuria (2001, 2002)	India	Emerging	Plant level	1975-76, 1988-89	Frontier	(?)	–
Barrios and Strobl (2002)	Spain	Developed	Plant level	1990-98	TFP	(?)	–
Smarzynska (2002)	Lithuania	Transition	Plant level	1996-2000	Prod function	(?)	(+)
Castellani and Zanfei (2003)	FR,IT, SP	Developed	Plant level	1993-97	TFP	SP (+), IT (-) FR (?)	–
Keller and Yeaple (2003)	US	Developed	Plant level	1987-96	TFP	(+)	–
Yudaeva et al. (2003)	Russia	Transition	Plant level	1993-97	Prod function	(+)	(-)
Javorcik and Spatareanu (2003)	Romania	Transition	Plant level	1992-2000	TFP	(+)	(-)
Damijan et al. (2003)a	8 CEE countries	Transition	Plant level	1994-98	Prod function	RO (-); other countries (?)	–

Table 2.1 continued   Main empirical findings on FDI externalities

Panel Data Studies							
Authors	**Host economy**	**Type**	**Data**	**Year**	**Dependent Variable**	**Intra-Industry Externalities**	**Backward Linkages**
Damijan et al. (2003)b	10 CEE countries	Transition	Plant level	Mid 1990s	Prod function	CZ,PL, RO, SK (+); Other countries (?)	–
Ruane and Ugur (2005)	Ireland	Developed	Plant level	1991-98	VA/L	(?)	–
Sabirianova et al. (2004)	Czech Republic and Russia	Transition	Plant level	1992-2000	Frontier	(-)	–
Driffield (2004)	UK	Developed	Industry	1984-97	Prod function	(-)	–
Torlak (2004)	5 EEC Countries	Transition	Plant level	1993-99	Prod function	CZ, RO (-); HU (+); PO, BL (?)	–
Karpaty and Lundberg (2004)	Sweden	Developed	Plant level	1990-2000	TFP	(+)	
Blyde et al. (2004)	Venezuela	Developing	Plant level	1995-2000	TFP	(?)	(?)
Lutz and Talavera (2004)	Ukraine	Transition	Plant level	1998-99	VA/L	(+)	–
Sinani and Meyer (2004)	Estonia	Transition	Plant level	1994-99	Prod function	(+)	–
Javorcik et al. (2004)	Romania	Transition	Plant level	1998-2000	Prod function	(+)	(+)
Barry et al. (2005)	Ireland	Developed	Plant level	1990-98	Prod function	(-)	–
Girma (2005)	UK	Developed	Plant level	1989-99	TFP	(+)	–
Taki (2005)	Indonesia	Developing	Plant level	1990-95	Prod function	(+)	–
Barrios et al. (2006)	Ireland	Developed	Plant level	1990-98	TFP	(+)	–
De Propris and Driffield (2006)	UK	Developed	Industry	1993-98	Prod function	(+)	(+)
Békés et al. (2006)	Hungary	Transition	Plant level	1992-2003	TFP	(+)	(?)

Table 2.1 continued   **Main empirical findings on FDI externalities**

Panel Data Studies							
Authors	Host economy	Type	Data	Year	Dependent Variable	Intra-Industry Externalities	Backward Linkages
Driffield (2006)	UK	Developed	Industry	1984-92	Prod function	(+)	(-)
Girma and Wakelin (2007)	UK	Developed	Plant level	1980-92	TFP	(+)	(+)
Girma and Görg (2007)	UK	Developed	Plant level	1980-92	TFP	(+)	
Crespo et al. (2007)	Portugal	Developed	Plant level	1996-2000	VA/L	(?)	(?)
Haskel et al. (2007)	UK	Developed	Plant level	1972-92	Prod function	(+)	–
Abraham et al. (2007)	China	Emerging	Plant level	2002-04	TFP	(+)	–
Liu (2008)	China	Emerging	Plant level	1995-2000	TFP	(?)	(+)
Girma et al. (2008)	UK	Developed	Plant level	1992-99	TFP	(+)	(+)
Blalock and Gertler (2008)	Indonesia	Developing	Plant level	1988-96	Prod function	(?)	(+)

Notes: (+) = positive FDI externalities, (-) = negative FDI externalities, (?) = insignificant, - = not estimated.

Several studies present empirical findings from a variety of specifications of estimated regression models. The results listed in the table represent the findings from specifications that are either clearly preferred by the author(s) or appear to offer the most robust empirical findings among the set of presented results.

There are four different estimation strategies. VA/L indicates the estimation of determinants of labour productivity of domestic firms measured by the ratio of value added over number of employees. Frontier indicates studies that use the level of deviation of productivity of domestic firms to best practice firms as dependent variable. Production function represents studies that estimate the productivity effect of FDI within a production function framework. TFP indicates studies that first obtain a score of total factor productivity (TFP) for domestic firms, which is then regressed on industry FDI presence and other control variables.

Country abbreviations: BL=Bulgaria, RO=Romania, P=Poland, SP=Spain, IT=Italy, FR=France, CZ=Czech Republic, SK=Slovakia, HU=Hungary, CEE=Central and Eastern Europe.

The second wave of empirical research on FDI externalities benefitted from the growing availability of plant level datasets containing observations for several years, which has led to the development of several strategies to identify FDI spillovers. For instance, several studies estimate the externality effect from industry FDI presence on the level of efficiency deviation of domestic firms relative to best practice firms. Other studies estimate productivity effects from FDI within the framework of production functions, either directly or via the effect of FDI on total factor productivity scores of domestic firms. The findings from these panel data studies challenge the original positive findings from cross-sectional studies. Whereas the early cross-sectional studies tend to present evidence of positive intra-industry FDI externalities, findings from panel data studies are characterised by a substantial level of heterogeneity. Previous surveys have linked this difference between cross-sectional evidence of positive FDI effects and findings of insignificant or negative intra-industry FDI effects from several panel data studies to differences in the nature of the data that is used for the empirical estimations (e.g. Görg and Strobl 2001, Hanson 2001b). However, looking at the variety of findings presented by the most recent panel data studies, this dichotomy in findings is no longer present in the evidence. It is true that several panel data studies have produced evidence of insignificant FDI effects (e.g. Barrios 2000, Kinoshita 2001, Girma et al. 2001) or significant negative FDI effects (e.g. Aitken and Harrison 1999, Djankov and Hoekman 2000, Konings 2000). At the same time, however, several other panel data studies have generated evidence of significant positive FDI effects (e.g. Haskel et al. 2007, Yudaeva et al. 2003, Taki 2005).[4]

Overall, it is clear that the empirical findings from more recent studies counter the original optimism that positive intra-industry FDI externalities are a common phenomenon. At the same time, the level of heterogeneity is too large to argue that panel data studies favour any type of estimated FDI effect. A good example of the extent of heterogeneity in findings is indicated by the findings from several studies on FDI effects in the host economy Romania. Konings (2000) and Torlak (2004) both present plant-level based evidence of negative FDI externalities in Romania for the 1990s. Damijan et al. (2003a, 2003b) find evidence of positive FDI externalities in Romania in their first study and evidence of negative externalities in their second study. Javorcik and Spatareanu (2003) and Javorcik et al. (2004) present quantitative evidence that strongly indicates the presence of positive intra-

---

4   The difference in findings from cross-sectional studies and the initial panel data studies fuelled the suspicion that cross-sectional evidence of positive FDI externalities may be affected by several econometrical problems. Most importantly, the critique of cross-sectional studies entails that there are problems of causality or endogeneity concerning the variable measuring the industry level of foreign participation (Hanson 2001b, Aitken and Harrison 1999, Rodrik 1999), causing cross-sectional studies to produce an upward bias in their estimations of FDI spillovers. I discuss this issue at length in Chapter 5, where I conduct cross-sectional estimations of intra-industry FDI externalities in Mexican manufacturing industries.

industry FDI externalities among Romanian manufacturing plants. These marked differences in findings for the same host economy indicate that differences in datasets, methodology and econometric choices and specifications may have far-reaching effects on the empirical findings.

FDI Externalities through Backward Linkages

In response to the markedly high level of heterogeneity of the evidence, recent research is trying to improve the identification of FDI externalities. An important development is that there appears to be a growing recognition that FDI spillovers may occur both within and between industries. In fact, one could argue that inter-industry externalities are more likely to arise than intra-industry externalities. One reason for this is that FDI firms will try to protect their ownership-specific advantages from their competitors, thereby minimising positive intra-industry externalities (Kugler 2006; Moran 2005). In contrast, input-output linkages between FDI and local suppliers are more likely to foster externality effects, as markets for inputs are characterised by the frequent transmission of information and knowledge between buying and selling parties (Lall 1980) and because foreign-owned firms are often involved in the provision of several types of support to their suppliers.

Despite the growing recognition that inter-industry FDI effects may be important, Table 2.1 clearly indicates that only a minority of empirical studies contains an estimation of externalities from FDI to their suppliers in host economies.[5] A well-known panel data study that presents evidence of positive FDI externalities among domestic suppliers is Blalock and Gertler (2008). They regress productivity of Indonesian manufacturing plants on a vector of control variables, including intra-industry FDI participation and FDI participation in input-supplying industries (based on input-output tables). This inter-industry FDI participation variable carries a significant positive coefficient, indicating the presence of positive FDI externalities through backward linkages, whereas the estimated effect of intra-industry foreign participation is insignificant. A related study is presented by Smarzynska (2002), who finds that only inter-industry FDI participation generates positive FDI spillovers through input-output linkages among domestic manufacturing plants in Lithuania. Additional evidence of positive spillovers that occur between industries is presented for the UK by Driffield et al. (2004) and Girma and Wakelin (2007), for China by Liu (2008) and for Romania by Javorcik et al. (2004).

Having said so, it is clear that the evidence on FDI spillovers through backward linkages is also characterised by a substantial level of heterogeneity. Blyde et al. (2004) for instance present the findings from a large number of different

---

5   In line with the evidence, I confine the discussion here to spillovers through backward linkages between FDI and their suppliers. There are a limited number of studies that include effects through forward linkages between FDI firms and domestic client firms, see e.g. Driffield et al. (2004).

estimations of FDI externalities among Venezuelan manufacturing plants, finding no evidence of externality effects through input-output linkages. Similar findings of an insignificant inter-industry FDI effect are presented by Békés et al. (2006) for Hungary and Crespo et al. (2007) for Portugal. Furthermore, Driffield (2006), Javorcik and Spatareanu (2003) and Yudaeva et al. (2003) present findings that indicate a negative association between the inter-industry presence of FDI and productivity of domestic plants, suggesting that these input-output linkages may also transmit negative externalities. Recalling the discussion on the concept of externalities in section 2, it is important to recognise that positive inter-industry FDI externalities only occur when productivity advantages obtained by domestic suppliers outweigh the benefits that FDI firms obtain (e.g. lower prices, shorter delivery times, etc.). The empirical findings of a negative association between inter-industry foreign participation and productivity of domestic suppliers suggest that there are cases where FDI firms obtain more benefits than suppliers do from inter-industry input-output linkages, resulting in the creation of negative externalities.

4. Absorptive Capacity and the Technology Gap

The heterogeneous nature of the evidence, in combination with empirical findings from macro-oriented studies on FDI growth effects that indicate that there appear to be several structural factors that may be important for these growth effects to arise, has spurred micro-economics research to indentify factors that facilitate or maximise FDI externality effects (Blomström and Kokko 2003, Crespo et al. 2007). For instance, Driffield and Love (2007) analyse industry level data for the UK and find that the underlying motivation of FDI firms in the UK appears to be related to the generation of positive externalities (see Driffield and Love 2007). Another factor may be the nationality of FDI, as several studies identify structural differences in the generation of externalities between FDI firms from different home countries (Haskel et al. 2007, Javorcik et al. 2004, Girma and Wakelin 2007). Also, the level of foreign participation in FDI firms may be important (Sjöholm and Blomström 1999, Dimelis and Louri 2002, Abraham et al. 2007, Javorcik and Spatareanu 2003). Although research on such determinants of FDI externalities appears to be promising, the quantity of the evidence has remained limited thus far.

Evidence on the importance of the level of absorptive capacity of domestic firms for the materialisation of FDI spillovers is more plentiful. The main idea of the concept of absorptive capacity is that only those domestic firms or industries that possess a sufficient level of absorptive capacity are able to absorb new knowledge and technologies from FDI. Table 2.2 lists the findings from empirical studies that have estimated the effect of absorptive capacity on FDI externalities. Crucially, as there is no direct measure of absorptive capacity, researchers resort to the use of proxy indicators. One such proxy is the level of R&D spending by domestic firms, where a high level of R&D expenditure is equated with a high level of absorptive capacity. Both Kinoshita (2001) and Kathuria (2001, 2002) find that positive FDI externalities are facilitated among domestic firms with a relative high level of R&D spending, which they interpret as evidence that FDI effects are positively influenced by a high level of absorptive capacity. Another approach is to relate the size of domestic firms to the level of absorptive capacity. For instance, both Blyde et al. (2004) and Aitken and Harrison (1999) find that small domestic firms are subject to negative FDI externalities, which they interpret as evidence that a limited level of absorptive capacity makes these domestic firms less able to compete with foreign-owned firms, leading to negative FDI effects. Furthermore, Blyde et al. (2004) find positive FDI externalities among large domestic firms, suggesting that these firms have sufficient absorptive capacity to absorb new technologies from FDI or to become more efficient in response to the increased level of competition caused by the entrance of FDI firms. The use of the export status of domestic firms as indicator of absorptive capacity appears to be more problematic: whereas Barrios and Strobl (2002) find that positive FDI externalities are more pronounced among exporting firms, Abraham et al. (2007) and Barrios (2000) find that it is exporting firms that are subject to negative FDI effects.

The majority of studies listed in Table 2.2 relate the concept of absorptive capacity to the level of technological differences between FDI and domestic firms. This interpretation harks back to the catch up thesis on the importance of international flows of technologies and knowledge between advanced and lagging countries as stimulus of economic growth of the latter group of countries (Abramovitz 1986, Gershenkron 1962, see also Keller 1996, Cohen and Levinthal 1989, 1990). FDI studies that adopt this catch up thesis interpret the technology gap between FDI and domestic firms as direct inverse indicator of the level of absorptive capacity of domestic firms. An example of corroborating evidence is Haddad and Harrison (1993), who find that positive FDI externalities only appear to materialise in low technology industries in Morocco. Under the assumption that the level of technological differences between FDI and domestic firms is relatively low in low technology industries, this finding supports the interpretation of the technology gap as inverse indicator of the level of absorptive capacity of domestic firms. Girma et al. (2001) and Imbriani and Reganati (1997) find that a large technology gap stimulates negative FDI spillovers, suggesting that firms with a low level of absorptive capacity are unable to compete with FDI firms, resulting in negative competition induced pecuniary externalities. Finally, Girma (2005), Kokko (1996) and Taki (2005) all find that positive FDI externalities are facilitated by a small technology gap between FDI and domestic firms.

Despite this corroborating evidence, it is important to recognise that equating the size of the technology gap with the inverse level of absorptive capacity of domestic firms reflects the underlying catch up thesis only partially. It is true that an important element of the original catch up thesis emphasises that lagging countries need to possess a sufficient level of absorptive capacity to be able to absorb new technologies. Furthermore, however, the catch up effect is *positively* related to the size of technological differences between countries, as the technology

Table 2.2   Empirical findings on the importance of absorptive capacity for FDI externalities

Authors	Country	Absorptive Capacity	Effect on FDI Externalities
Haddad and Harrison (1993)	Morocco	Low tech versus high tech industries	Positive FDI externalities in low tech industries
Kokko (1994)	Mexico	Interaction variable FDI and Technology Gap	Positive FDI externalities do not arise in industries with high FDI and high technology gap
Blomström and Wolff (1994)	Mexico	Technology Gap	Technology Gap positive effect on FDI externalities
Kokko et al. (1996)	Uruguay	Technology Gap	Positive FDI externalities only when technology gap is small
Imbriani and Reganati (1997)	Italy	Technology Gap	Positive FDI externalities in low technology gap industries, negative FDI externalities in high technology gap industries
Aitken and Harrison (1999)	Venezuela	Size of firm	Negative FDI externalities among small firms
Sjöholm (1999)	Indonesia	Technology Gap	FDI externalities promoted by large technology gap
Barrios (2000)	Spain	Plant level R&D spending; Export Status	Low R&D industries negative FDI externalities; negative externalities among exporting domestic firms
Zukowska-Gagelmann (2000)	Poland	Interaction between Technology Gap and FDI	High FDI with high Technology Gap promotes positive FDI externalities
Girma et al. (2001)	UK	Technology Gap	High technology gap lowers FDI externalities
Kathuria (2001, 2002)	Indonesia	R&D, Technology Gap	positive effect of technology gap on FDI externalities in scientific industries; negative effect on FDI externalities in non-scientific industries
Kinoshita (2001)	Czech Republic	R&D spending	Positive FDI externalities among Czech firms that are R&D intensive
Barrios and Strobl (2002)	Spain	Exporting status	Positive FDI externalities among exporting domestic firms
Castallani and Zanfei (2003)	Spain, Italy, France	Interaction TFP gap * FDI	High TFP gap with high FDI presence promotes positive spillovers
Karpaty and Lundberg (2004)	Sweden	Interaction variable R&D and industry FDI	Positive effect of interaction variable on FDI externalities
Blyde et al. (2004)	Venezuela	Size of firm	Negative FDI externalities among small/ medium firms; positive FDI externalities among large firms
Sinani and Meyer (2004)	Estonia	Age of firm	Positive FDI externalities among young firms; old firms no FDI externalities

Table 2.2 continued   **Empirical findings on the importance of absorptive capacity for FDI externalities**

Authors	Country	Absorptive Capacity	Effect on FDI Externalities
Girma (2005)	UK	Technology Gap	Positive FDI externalities when technology gap is small
Taki (2005)	Indonesia	Technology Gap	Positive FDI externalities larger in low technology gap industries
Svejnar et al. (2007)	17 EEC countries	Age of firm	Positive FDI externalities only among old firms
Girma and Görg (2007)	UK	Interaction variable FDI and Technology gap	Non linear, decreasing technology gap, first negative spillovers, then positive spillovers
Haskel et al. (2007)	UK	Technology Gap	Technology Gap positive effect on FDI externalities
Abraham et al. (2007)	China	Export status domestic firms	Exporting domestic firms subject to negative FDI externalities
Girma et al. (2008)	UK	Export status domestic firms	Positive FDI externalities larger among exporting domestic firms

gap indicates the potential magnitude of the catch up effect (Abramovitz 1986, Gershenkron 1962). If there are large technological differences, there will be a large growth potential for lagging countries, provided that they posses a sufficient level of absorptive capacity. This means that the interpretation of the technology gap between FDI and domestic firms as direct inverse indicator of the level of absorptive capacity is flawed, as it fails to appreciate the importance of the presence of sufficiently large growth opportunities in the process of technological catch up.

Therefore, one could argue alternatively that large technological differences between FDI and domestic firms may be conducive rather than detrimental for the occurrence of positive FDI externalities. One argument in support of this alternative interpretation is that the size of the technology gap indicates the potential scope for positive externalities (Findlay 1978, Blomström and Wang 2002, Jordaan 2005, 2008c). Furthermore, the presence of a large potential of positive FDI externalities may stimulate domestic firms to engage in externality facilitating investments. Following Rodrik (1992), one could interpret the occurrence of positive externalities as the outcome of an investment decision. Domestic firms are more likely to make externality-facilitating investments when the potential gains are large. In other words, domestic firms will try to improve their level of absorptive capacity when the level of technological differences between FDI and domestic firms is sufficiently large. Finally, there is also the argument that, under the condition of large technological differences, the negative competition effect is less likely to arise, as FDI and domestic firms are most likely to be in direct competition when they are technologically similar.

Table 2.2 shows that several studies present findings that are in direct contrast to the traditional interpretation of the technology gap and more in line with the alternative interpretation as described above. For instance, Castellani and Zanfei (2003) find a positive effect of an interaction variable between industry level FDI and the level of technological differences between domestic and foreign-owned firms, indicating that industries with a large presence of FDI and a large technology gap are subject to positive FDI externalities. A similar effect of such an interaction variable is presented by Zukowska-Gagelmann (2000) for FDI externalities among Polish manufacturing plants. Furthermore, an estimated direct positive effect of the technology gap on positive FDI spillovers is presented by Haskel et al. (2007) for the UK, Sjöholm (1999) for Indonesia, Blomström and Wolff (1994) for Mexico and Kathuria (2001, 2002) for India.

These findings of a positive effect of large technological differences on positive FDI externalities challenge the interpretation of the technology gap as direct inverse indicator of the level of absorptive capacity of domestic firms. Having said so, studies that have produced evidence of a positive effect of large technological differences on positive FDI externalities offer little explanation for this finding, usually confining interpretations to statements that this evidence is not in support of the absorptive capacity thesis. As I have discussed above, there are several reasons why large technological differences may actually have such a positive effect on FDI externalities. Furthermore, whether the technology gap has a negative or positive effect on FDI externalities has far-reaching policy implications. Following the notion that technological differences should be limited in order for positive externalities to arise, host economy governments should aim to attract new FDI firms that are not too dissimilar from domestic firms. In contrast, following the idea that large technological differences foster positive FDI externality effects leads to policies to attract FDI firms that are technologically sufficiently advanced from domestic firms. Clearly, the incompleteness of the link between the technology gap as inverse indicator of absorptive capacity with the underlying catch up thesis, the contrasting evidence on the effect of the technology gap on FDI spillovers and the far-reaching policy implications that are attached to these findings call for further empirical investigation of the effect of the technology gap on FDI spillovers.

5. Spatial Dimensions of FDI Externalities

Within the context of recent attempts to obtain a better identification of the full range of FDI externalities and to establish a better understanding of when these effects arise, several studies have started to explore spatial dimensions of FDI effects. Although most research that incorporates spatial dimensions of FDI spillovers does not link explicitly into the broad literature on relations between externalities and geographical space, this literature contains several indications that spatial dimensions are very likely to be important for FDI externality effects. For instance, an impressive body of research has developed on place-based externalities that are directly linked to the existence of agglomerations of economic activity (for surveys see Moomaw 1981, Eberts and McMillen 1999, Rosenthal and Strange 2004). The underlying mechanisms of these agglomeration economies include thick labour markets, the establishment of specialised suppliers of (non-traded) inputs and the build up of local pools of knowledge which generate knowledge spillovers (Gordon and McCann 2000, Duranton and Puga 2004, Henderson 2001). Importantly, these mechanisms are very similar to the channels that generate and transmit FDI externality effects (Jordaan 2005). For both types of externality, inter-firm labour mobility, inter-firm input-output linkages and knowledge spillovers (demonstration effects) are the channels that facilitate these productivity effects. This large degree of similarity suggests that channels of FDI spillovers are more likely to come into existence and be more effective when FDI firms operate in agglomerations of economic activity within host economies. In light of this, the hypothesis that FDI spillovers will be facilitated when foreign-owned firms operate in agglomerations of economic activity is easily put forward.

Another line of research that incorporates the analysis of relations between geographical space and externalities focuses on the important role that geographical proximity can play in the generation and transmission of knowledge spillovers (see Audretsch and Feldman 2004, Henderson 2001, also Döring and Schellenbach 2006). Using the concept of the knowledge production function, empirical studies using country and industry level data have been successful in identifying relationships between knowledge inputs and outputs (see Grilliches 1979, 1992). The evidence of such relationships at the micro-level is much less convincing. This feature that the knowledge production function is more successfully identified at more aggregate levels of activity strongly indicates the presence of externalities, which are not identified when using firm level data on innovation inputs and outputs (Audretsch and Feldman 2004). Research using regionalised versions of the knowledge production function has produced important indications of the presence of regionally confined externalities, in the form of evidence that innovations are positively affected by both regional university and corporate R&D (e.g. Jaffe 1989, Acs, Audretsch and Feldman 1994, see also Audretsch and Feldman 2004). Related findings are presented by Jaffe et al. (1993) and Jaffe and Trajtenberg (2002), who identify the importance of geographical proximity between economic agents for knowledge spillovers in the form on significant levels of localisation of knowledge creation in the US. Also, there is of course the research field on human capital externalities, which provides substantive evidence of productivity differences between regions or cities that are linked to regionally-confined human capital externalities (see Moretti 2004).

The relevance of these findings on the importance of geographical proximity for the generation and transmission of knowledge spillovers for estimations of FDI externalities is two-fold. First, the recognition that geographical proximity fosters knowledge spillovers underlines the likelihood that FDI spillovers are facilitated when FDI firms locate in agglomerations of activity within host economies.

Second, the importance of regionalising the knowledge production function to capture regionally confined externalities from knowledge creation suggests that the estimation of FDI spillovers will also benefit from incorporating such a regional dimension to these FDI effects. This means that, in addition to estimating general nation wide FDI productivity effects in a host economy, estimations should aim to identify separate FDI externalities at the regional level, as this will provide a better estimation of the range of FDI effects that may occur.

Finally, a third research field on the relation between geographical space and externalities is focusing on the identification and quantification of knowledge spillovers from research and development and other innovative activities across geographical space. Although geographical proximity is likely to facilitate the transmission of knowledge spillovers, this is not to say that these spillovers can not be transmitted between regions. In other words, although research on spatial dimensions of knowledge spillovers does acknowledge that it is very likely that geographical distance has a negative effect on the spatial reach of knowledge spillovers, this negative effect is not necessarily so strong that inter-regional spillovers do not occur. Recent empirical evidence on the occurrence of such inter-regional effects is presented by Anselin et al. (1997, 2000) for the US, Bode (2004) for Germany and Moreno et al. (2005), Crescenzi et al. (2007) and Rodríguez-Pose and Crescenzi (2008) for regions in the European Union. These studies present robust evidence that knowledge output in a given region is affected by knowledge inputs in other regions. At the same time, the findings also confirm the negative effect of distance on these spillover effects, as spatial knowledge spillovers are subject to negative spatial decay effects. Clearly, these findings are very relevant for the estimation of FDI spillovers, as they suggest that FDI effects may also materialise between regions in a host economy.

FDI Externalities and Geographical Space

Table 2.3 presents the main findings from recent research that incorporates some form of assessment of spatial dimensions of FDI externalities. One interpretation of the possibility that spatial dimensions of FDI externalities may be important translates into distinguishing between FDI spillovers at the national and the regional level. Several studies estimate a separate effect of regional FDI participation on productivity of a given domestic plant in a host economy, producing mixed findings. Aitken and Harrison (1999), Konings (2000), Yudaeva et al. (2003), Torlak (2004) and Haskel et al. (2007) present findings containing an estimated insignificant effect from regional foreign participation on productivity of domestic plants. In contrast to this, Karpaty and Lundberg (2004), Blyde et al. (2004), Girma and Wakelin (2007) and Halpern et al. (2007) present findings of a significant positive effect from regional foreign participation. This heterogeneity in findings indicates that there may be a regional dimension to FDI spillovers, but that such a regional dimension is far from automatic. Having said so, several studies use a variable capturing total regional foreign participation or control only for the effect of intra-

Table 2.3   Empirical findings on spatial dimensions of FDI externalities

Authors	Country	Spatial Dimensions of FDI	FDI Externalities
Aitken and Harrison (1999)	Venezuela	Intra-industry intra-regional	(?)
Sjöholm (1999)	Indonesia	Intra-industry intra-regional; inter-industry intra–regional	Intra-industry intra-regional (-) at district level; inter-industry intra-regional (+) at province and district level
Konings (2000)	Transition countries	Intra regional intra industry	(?)
Driffield and Munday (2001)	UK	FDI in agglomerated and non agglomerated industries	(+) only in agglomerated industries
Yudaeva et al. (2003)	Russia	Intra-industry intra-regional	(?)
Driffield (2004)	UK	Intra-industry intra regional; intra-industry inter-regional; inter-industry intra-regional	Intra-industry inter-regional (-); intra-industry intra-regional and inter-industry intra-regional (+)
Driffield et al. (2004)	UK	Variety of forward (F) and backward (B) linkages	F: inter-industry intra-regional and intra-industry inter-regional (+); B: intra industry inter-regional and inter-industry inter-regional (-)
Torlak (2004)	Transition countries	Intra-industry intra regional	(?)
Karpaty and Lundberg (2004)	Sweden	Intra-industry intra-regional	(+)
Blyde et al. (2004)	Venezuela	Intra-industry intra-regional	(+)
Smarzynska (2002)	Lithuania	Intra and inter regional intra and inter-industry	Intra- and inter-regional intra-industry (?), intra- and inter-regional inter-industry (+)
Girma (2005)	UK	Intra-industry intra-regional; intra-industry inter-regional	Intra-industry intra-regional (+); intra-industry inter-regional (+) in some estimations
De Propris and Driffield (2006)	UK	Intra-industry intra regional; intra-industry inter-regional; inter-industry intra-regional	Agglomerated industries: intra-industry intra-regional and inter-industry intra regional (+); Non agglomerated intra-industry intra-regional (-)
Driffield (2006)	UK	Intra and inter regional, intra and inter-industry	Intra industry intra regional and inter-industry intra regional (+), other (?)

Table 2.3 continued   **Empirical findings on spatial dimensions of FDI externalities**

Authors	Country	Spatial Dimensions of FDI	FDI Externalities
Barrios et al. (2006)	Ireland	Intra industry intra regional	(+) in industries that are agglomerated
Girma and Wakelin (2007)	UK	Intra industry intra regional, intra-industry inter-regional; inter-industry intra regional	Intra industry intra regional (+), intra industry inter-regional (?), inter-industry intra regional (+)
Crespo et al. (2007)	Portugal	Intra industry intra-regional; forward and backward intra-regional	Intra industry intra regional (-), backward linkages intra-regional (+)
Halpern et al. (2007)	Hungary	Intra and inter regional intra and inter-industry	Intra industry intra regional (+); inter industry inter regional (+), other (?)
Abraham et al. (2007)	China	Intra-industry intra-regional	(+)
Haskel et al. (2007)	UK	Intra industry intra regional	(?)
Blalock and Gertler (2008)	Indonesia	Intra-industry intra-regional, inter industry intra-regional	Inter-industry intra-regional (+), intra industry intra-regional (?)

Notes: (+) = positive FDI externalities, (-) = negative FDI externalities, (?) = insignificant.

There are four possible types of FDI participation in relation to regional FDI effects: intra-regional intra-industry, intra-regional inter-industry, inter-regional intra-industry and inter-regional inter-industry. Most studies do not include all four types in their estimated regression models, however. Table 2.3 indicates for each study which types of foreign participation are estimated.

industry regional foreign participation. This lumping together of intra- and inter-industry regional foreign participation or the omission of a variable that controls for the separate effect from inter-industry regional foreign participation may have affected the estimated regionalism of FDI spillover effects.

An example of a study that does distinguish between the effects of intra- and inter-industry regional foreign participation is Sjöholm's (1999) study of determinants of productivity of Indonesian manufacturing plants. The findings show that the estimated effect from intra-industry regional foreign participation is positive or negative, depending on the geographical scale of the region. The estimated positive effect from inter-industry regional foreign participation is positive and robust to changes in regional units. Driffield (2004, 2006) extends on this approach and estimates FDI effects among UK manufacturing industries, also distinguishing between intra- and inter-industry regional foreign participation.

The findings for the UK indicate that both types of regional foreign participation appear to generate positive externalities. Other studies that have estimated the effect of one or more of such regional FDI variables include Girma and Wakelin (2007), Smarzynska (2002), Crespo et al. (2007) and Blalock and Gertler (2008).

Another assessment of the spatial dimensions of FDI externalities is to address the effect of industry agglomeration or geographical proximity between FDI and domestic firms on FDI spillovers. As discussed earlier, the large extent of similarity between the mechanisms underlying agglomeration economies and the channels that generate and transmit FDI spillovers strongly suggests that FDI externalities are facilitated in agglomerations of economic activity. Despite this, only a very few empirical studies have actually estimated the direct effect of industry agglomeration or geographical proximity on FDI externalities. The findings from these studies indicate that agglomeration does indeed appear to have a positive effect on FDI spillovers. For instance, De Propris and Driffield (2006) find that positive FDI externalities only materialise in those UK manufacturing industries that are characterised by a relative high level of agglomeration. In strong contrast to this, their findings also indicate that non-agglomerated industries are subject to negative FDI effects. Barrios et al. (2006) calculate the level of co-agglomeration of Irish and foreign-owned manufacturing firms, finding that in a number of industries this feature of co-agglomeration is important. Furthermore, their estimations of productivity spillovers among Irish firms indicate that positive FDI effects only materialise in those industries that are characterised by this co-agglomeration.

Finally, some tentative attempts have been made to identify more explicitly the relation between geographical distance and FDI spillovers, by incorporating interregional FDI effects in estimations of FDI externalities. For instance, in estimating determinants of productivity among manufacturing plants in Lithuania, Smarzynska (2002) distinguishes between intra- and inter-industry foreign participation, at both the intra- and inter-regional level. The empirical findings indicate that there is an inter-regional dimension of FDI effects, as the variable capturing inter-regional inter-industry foreign participation carries a significant positive coefficient. Another example is Driffield (2006), who distinguishes between the same four types of FDI participation in UK regions. The findings suggest the absence of inter-regional effects, as only the intra-regional FDI variables carry significant coefficients. Having said so, in a related study on the UK, Driffield et al. (2004) find that foreign-owned firms appear to generate a variety of both intra- and inter-regional FDI effects. Girma (2005) finds some evidence of positive inter-regional intra-industry externalities among UK manufacturing plants. Finally, Halpern et al. (2007) conduct a study on FDI externalities in Hungary and present findings of positive spatial FDI externalities of an inter-industry nature.

Although several studies present findings of spatial FDI effects, the evidence needs to be interpreted with caution. One issue is that not all studies control for all types of intra- and inter-regional FDI participation. This introduces the problem of omitted variable bias, which may affect the estimated effect of the foreign participation variables that are included in the estimations. Furthermore, it appears that most studies that include an estimation of inter-regional FDI effects are based on ad-hoc specifications of the relation between geographical space and FDI spillovers. To identify correctly such spatial effects, two key criteria have to be met (Boden 2004, Jordaan 2008b). First, the estimation needs to capture the decay effect that geographical distance is likely to have on spatial spillovers. Importantly, as it is not possible to determine the relation between distance and spatial effects *a priori* (Anselin 1988), empirical estimations of spatial FDI effects should experiment with several specifications of this relationship. The need to do so becomes more apparent when considering that there are several channels that may generate and transmit FDI spillover effects. As the geographical reach of the effects from these different channels may be affected by geographical distance differently, it is important to experiment with several specifications of the relation between FDI spillovers and distance to assess whether spatial effects occur.

Therefore, studies that estimate spatial FDI effects based on only one specified relation between these spatial effects and geographical distance may not identify the entire range of spatial FDI externalities. The study on FDI effects among UK manufacturing plants by Girma and Wakelin (2007) provides a good example. Girma and Wakelin relate the level of intra-industry foreign participation in all UK regions to productivity of a UK manufacturing plant in a given region. To capture the effect of inter-regional distance on the process of spatial FDI spillovers, they weigh the regional foreign participation variable with distance in kilometres between these regions, which is an often-used approach to capture the effect of distance on spatial effects (Anselin 1988, Adserá 2000). In this specification, for a firm in the given region, the effect of inter-regional foreign participation depends positively on the size of the potential spatial FDI effects from other regions and negatively on the distance to these other regions. Girma and Wakelin (2007) find no significant effect of this spatially weighted inter-regional FDI variable, leading them to conclude that there are no spatial FDI effects. However, their use of inter-regional distances to capture the effect of geographical space on spillovers is based on the assumption that there is a continuous negative relation between distance and spillovers. Alternatively, it could be the case that the negative effect of distance is such that spatial spillovers are subject to a geographical cut off point, after which they seize to materialise. If this alternative relationship between spillovers and distance exists, it may be that the use of inter-regional distances as weighting variable makes that the estimation does not identify spatial spillovers that are subject to such a geographical cut off point.

The second criterion that estimations of spatial effects need to meet is the use of an appropriate measure of the scope of potential inter-regional FDI spillovers. Again, most studies have not fully considered the importance of this issue. Usually, in measuring the potential scope of inter-regional FDI effects, studies rely on spatially-weighted foreign participation variables that are simply the variables measuring intra- and inter-industry foreign participation at the intra-regional level. For instance, intra-industry intra-regional FDI participation is measured as the share of FDI firms in employment of a regional industry. The variable representing the scope of inter-regional intra-industry FDI effects between regions is then simply the (spatially weighted) share of FDI in regional industry employment in the regions. The main problem with this specification is that it may not reflect accurately the actual scope of inter-regional FDI effects. For instance, in the case where a small number of FDI firms have a large share in employment of a small regional industry, the commonly used measurement would indicate that the scope of potential inter-regional FDI effects from this region is large. Clearly, this is not in line with the small absolute volume of FDI in this region. Similarly, if a considerable number of FDI firms operate in a large regional industry, the standard measure would indicate that the scope of inter-regional FDI effects from this region is relatively small, whereas the absolute size of FDI in this region suggests otherwise. Therefore, as the commonly used indicators of inter-regional FDI effects may not accurately capture the magnitude of these effects, it may be the case that alternative measures that do incorporate the scale of regional FDI will produce more reliable results.

6. Summary and Conclusions

The last two decades have witnessed a rapidly growing interest in FDI effects that arise in the form of productivity effects among domestic firms in host economies. Competition effects, inter-firm labour mobility, input-output linkages and knowledge spillovers constitute the main channels through which these effects are transmitted. It appears to be incorrect to assume that these effects consist solely of technological externalities, as a closer examination shows that FDI firms can also be linked to pecuniary externalities. In essence, technological externalities occur when FDI firms act as source of new knowledge and technologies, whereas pecuniary externalities arise when productivity changes of domestic firms are caused by these firms changing their conduct in response to the presence and operations of foreign-owned firms.

The growing interest in FDI externalities has promoted the development of a large body of empirical quantitative research on these effects. The survey in this chapter identifies several important features of this research. First, there is a strong discussion on the prevalence of intra-industry FDI externalities. Whereas some argue that there is substantial evidence of such externality effects, others disagree and argue that the evidence is weak. Evidence of negative intra-industry FDI externalities further indicates the complexity of this issue. Previous surveys have pointed out the dichotomy in findings of positive externalities from cross-sectional studies versus findings of insignificant or negative externalities from panel data studies. As the survey in this chapter shows, such a clear dichotomy in findings is no longer apparent, as several recent panel data studies present evidence of positive intra-industry externalities, underlining the heterogeneous nature of the evidence on this type of FDI effect. In response to this, a number of studies distinguish between intra- and inter-industry FDI effects. Not only is there no reason to assume that FDI spillovers are confined at the intra-industry level, there are arguments that suggest that positive inter-industry externalities may be more likely to arise, in particular among domestic suppliers of FDI firms. However, although several studies present evidence of such positive inter-industry effects, other studies find that inter-industry FDI participation can also generate significant negative externalities among domestic suppliers.

Another important issue is the development of research on determinants of FDI externalities. In line with empirical research on the conditionality of FDI growth effects in a macro-economic setting, micro-economics research on FDI effects has started to estimate the effect of several firm and industry characteristics. The most commonly accepted factor is the concept of absorptive capacity, which entails that only those domestic firms that possess a sufficient level of absorptive capacity will be able to absorb positive FDI externalities or prevent the occurrence of negative FDI effects. Although this notion that the absorptive capacity is important is undisputed, the survey identifies important problems with the empirical translation of this concept. In particular, there are problems with using the technology gap between FDI and domestic firms as direct inverse indicator of the level of absorptive capacity of these domestic firms, as this empirical translation does not fully capture the underlying catch up thesis. Also, several studies present evidence that a large technology gap is conducive rather than detrimental for the occurrence of positive FDI externalities. As I argue in this chapter, the positive effect of the technology gap can be explained by the fact that large technological differences indicate that the potential of positive externality effects is large. This offers domestic firms incentives to engage in externality-facilitating investments to improve their level of absorptive capacity. Also, negative competition effects are less likely to occur when the technology gap is large. Given the limited attention that these problems with the use of the technology gap as determinant of FDI spillovers have received in the literature, coupled with the strongly contrasting policy implications of findings of a positive or negative effect of large technological differences on FDI externalities, it seems clear that more empirical research on the effect of the technology gap is called for.

In light of the high extent of heterogeneity of findings on both intra- and inter-industry FDI externalities and the problems that surround the effect of the technology gap, it is very important to continue to pursue new research avenues to obtain a better identification of the full range of FDI effects and to improve our understanding of the conditions that foster positive FDI externalities. Within this context, a number of recent studies have started to explore geographical dimensions of FDI externalities. One example of this is to identify the effect of agglomeration or geographical proximity on FDI spillovers, which is linked to the large degree of similarity between channels of FDI spillovers and mechanisms underlying agglomeration economies. This large degree of similarity fuels the strong suspicion that agglomeration will have a positive effect on FDI externalities. As the survey shows, a surprisingly small number of studies have investigated this hypothesis.

The findings of these studies are in support of the hypothesis that agglomeration enhances positive FDI spillovers. Another interpretation of the relation between geographical proximity and spillovers leads to the estimation of both national level and regional FDI effects. Following the premise that geographical proximity may influence mechanisms that generate and transmit spillovers, it may be the case that there are separate FDI effects at the national and regional level. The findings on this issue are diverse, indicating that the regional dimension of FDI externalities is far from automatic. Having said so, it appears that several studies suffer from problems of aggregation and/or omitted variable bias, suggesting that the evidence must be interpreted with caution.

The third interpretation of the relation between spillovers and geographical proximity has led to research into the extent and nature of FDI spillovers that occur across geographical space. Again, the findings are heterogeneous, but they contain important indications that such inter-regional FDI effects may be important. At the same time, the survey shows that several of these studies may suffer from omitted variable bias, as they tend not to control for all types of intra- and inter-regional foreign participation. Furthermore, these studies tend to rely on single specifications of the relation between geographical space and spatial spillovers, which, especially given the variety of channels that underlie FDI externalities, may result in a failure to identify spatial FDI effects. Also, there appear to be problems with the measurement of the scope of potential spatial FDI externalities. Clearly, the importance of geographical space in externality-generating and transmitting processes as shown in the broad literature on agglomeration economies and knowledge creation, in combination with the findings and challenges facing the initial set of studies that have incorporated assessments of geographical dimensions of FDI externalities, strongly suggest that more empirical research on these spatial dimensions will improve the identification and our understanding of the range of FDI externalities that can occur in host economies.

Chapter 3
Agglomeration and FDI Location in Mexican Regions

1. Introduction

Economic activity shows a persistent tendency to concentrate geographically in space. The explanation for this is that an agglomeration of firms generates positive externality effects that are uniquely linked to the existence of the agglomeration (Marshall 1890, Henderson 2001, 2007, Duranton and Puga 2004). These place-based externalities, or agglomeration economies, place firms in the agglomeration at a productivity advantage over firms located elsewhere. The channels that underlie these place-based externalities include search and match effects on input and labour markets (Henderson 2001, Gordon and McCann 2000, Duranton and Puga 2004), increasing levels of interconnectedness between and specialisation among firms (Kaldor 1970) and the creation of local pools of knowledge, fostering the generation and transmission of knowledge spillover effects (Henderson 2001, 2007, Rosenthal and Strange 2004). Although there are important debates on how to identify these agglomeration economies empirically, there is substantial evidence indicating that these place-based externalities can generate important productivity improvements among firms in such agglomerations (see e.g. Eberts and McMillen 1999, Rosenthal and Strange 2004, also Hanson 2001a).

Recent empirical studies of FDI location behaviour also present findings on the importance of agglomeration economies, as these studies estimate whether agglomeration economies are an important location factor for new FDI firms (e.g. Head et al. 1995, 1999, Barry et al. 2003, Crozet et al. 2004, Hilber and Voicu 2009). The appealing aspect of this analysis is that new FDI firms can be interpreted as being exogenous to the existing geographical distribution of economic activity in a host economy. Therefore, if new FDI firms are attracted to those regions in a host economy that contain agglomerations of economic activity, *ceteris paribus*, we can interpret this as evidence that agglomeration economies are an important location factor for these firms. The majority of these location studies present evidence that confirms that new FDI firms are indeed attracted to agglomerations of economic activity within host economies.

The relevance of these findings by recent FDI location studies for research on the operations and effects of FDI in Mexico is two-fold. First, they provide important indications of which regional characteristics may play an important role in the location process of new FDI activity within Mexico. The dramatic increase in the level of foreign participation in the Mexican economy in the last two decades has received considerable attention in the literature. Relatively little is known on which factors influence the location choice of new FDI firms within Mexico, however. Second, in the previous chapter I argued that the relation between agglomeration and FDI externalities may be important to analyse, an argument which is supported by findings of a positive association between the regional presence of an agglomeration of economic activity and the arrival of new FDI firms. Clearly, evidence that agglomeration economies play a similarly positive role in the location process of FDI in Mexico would constitute important support for addressing the relationship between agglomeration and FDI externalities in this host economy.

The chapter is constructed as follows. In section 2 I briefly review empirical evidence on FDI location factors. In section 3 I provide an overview of the main features of FDI in Mexico in the last 20 years, focusing on developments in the overall level of FDI, its sectoral distribution and developments of the maquiladora industries. Furthermore, I also present indicators of the regional distribution of FDI activity, against the background of locational changes of economic activity in Mexico following the introduction of economic liberalisation and trade promotion in the 1980s.

Section 4 discusses the econometrical model and data that I use to identify state characteristics that play the location behaviour of new FDI firms. I use a thus far unexplored dataset which contains information on the location decisions of a large set of new manufacturing firms for the period 1994-1999, a period which is characterised by a large increase in the number of FDI firms in Mexico. The nature of the dataset allows me to conduct conditional logit estimations, similar to recent studies on FDI location decisions by e.g. Head et al. (1995, 1999), Crozet et al. (2004) and Hilber and Voicu (2009). Importantly, in addition to calculating variables that control for state characteristics such as demand, labour costs, schooling and labour quality, I take great care in calculating several variables that capture the presence of agglomeration economies from a variety of sources.

Section 5 presents the main empirical findings of the analysis of FDI location behaviour. The evidence is particularly rich when it comes to identifying the effects of agglomeration economies, as I distinguish between agglomeration economies associated with the regional presence of manufacturing firms, financial services and commercial providers of material inputs. Furthermore, I distinguish between the regional presence of agglomerations of these activities of Mexican and foreign-owned firms separately. The analysis also addresses the question whether export oriented FDI firms attach different importance to location factors. Finally, the findings contain important evidence of the extent to which the effects of agglomeration economies are confined within regions, or alternatively are also transmitted spatially between regions. Finally, section 6 summarises and concludes.

2. FDI Location Factors: A Brief Survey

Several FDI location studies have produced important empirical evidence on which regional characteristics influence the regional distribution of new FDI manufacturing firms within host economies. Usually, four main types of location factor are distinguished: regional demand, regional production costs, regional policies that influence the attractiveness of regions for new FDI firms and the regional presence of agglomeration economies. Of these four main sets of location factors, it can be difficult to estimate the effect of public policies, as data on such policies is often simply unavailable or the regional variation of these policies is difficult to quantify.

Looking first at the findings on the importance of regional non-agglomeration related characteristics, there is evidence that a variety of these characteristics can be important. Several studies find that FDI firms prefer to locate in regions with high (potential) demand. For instance, Coughlin et al. (1991) find that the probability that a region is selected by a new FDI firm is positively affected by the regional level of income per capita. Similar findings are presented for related proxies of regional demand, such as population density (Woodward 1992), market size (Head et al. 1999) and overall regional income (Coughlin and Segev 2000).

A central element of regional production costs are the costs of the production factor labour. Here, the findings are more diverse. Studies such as Coughlin et al. (1991), Friedman et al. (1996), Canfei He (2003), Crozet et al. (2004) and Hilber and Voicu (2009) all find a negative effect of the regional wage level on the arrival of new FDI, suggesting that new FDI firms prefer to locate in regions with relative low wages. In contrast, Head et al. (1999) and Guimarães et al. (2000) find a positive effect of regional wages. The explanation for an estimated positive effect of wages is that wages also incorporate the productivity level of workers. This difference in the estimated effect of regional wages indicates that an empirical analysis of FDI location factors should try to distinguish between the cost and productivity elements that are incorporated simultaneously in the wage level.[1]

Most studies that do include an assessment of the effect of regional public policies find that these policies are important. Head et al. (1999) for instance find that regional corporate taxes, the absence of heterogeneity in regional business taxes and regional labour subsidies are all significantly associated with the probability that a region is selected by a new FDI plant. Taylor (1993) finds that the development status of a UK region influences the level of new FDI. Canfei He (2003) presents findings indicating that the regional existence of FDI policies influences the regional distribution of FDI within China. Basile et al. (2003) find

---

1   An alternative labour-related variable is the regional unemployment rate (see e.g. Woodward 1992, Head and Mayer 2004, Yamawaki 2006). Other regional cost variables include electricity (Carlton 1983) and the quantity or quality of road networks (Woodward 1992, Coughlin and Segev 2000, Yamawaki 2006).

that EU regions that are eligible for various types of support from the European Union have a higher probability to attract new FDI firms.

Turning to agglomeration economies, the evidence is based on the use of proxy variables, in the absence of direct indicators of these place-based externalities. Examples of such proxies are the regional number of manufacturing plants (e.g. Woodward 1992), the number of manufacturing employees or the share of the manufacturing sector in total regional employment (Carlton 1983, Coughlin and Segev 2000, Yamawaki 2006) or the level of manufacturing density (Coughlin et al. 1991, Friedman et al. 1996). The findings show a positive effect of these proxies on the probability that a region is selected by new FDI firms, which is interpreted as evidence that agglomeration economies are important as FDI location factor. However, the use of such proxy variables means that the evidence needs to be interpreted with caution, as they may alternatively simply capture the positive effect of the regional presence of a manufacturing culture on the arrival of new FDI firms. Also, the broad nature of the variables may lead to the problem that they capture the effect of omitted variables that influence FDI location decisions instead (Hanson 2001b).

In light of this, findings from studies that use several agglomeration variables may be more robust. In fact, several studies distinguish between the effects from agglomeration economies from domestic firms and from existing FDI plants on the attractiveness of a region for new FDI activity. For example, Head et al. (1995) find that US regions with a large number of US and Japanese manufacturing firms in the same industry of new Japanese firms have a higher probability to be selected by these new FDI firms. Hilber and Voicu (2009) present evidence indicating that the probability that a region is selected by new FDI firms in Romania is positively associated with the number of FDI and Romanian plants in the same industry of these new FDI firms and with the level of density of regional services. Other studies that use several agglomeration economies variables in regression models on FDI location decisions include Head et al. (1999) for the US, Crozet et al. (2004) for France, Guimarães et al. (2000) for Portugal, Canfei He (2003) for China and Basile et al. (2003) for the EU. These studies offer convincing evidence of the importance of agglomeration economies, as several of the agglomeration economies variables are positively associated with the probability that a region is selected by new FDI firms. One aspect where the findings differ concerns the relative importance of agglomeration economies from the various sources. In particular, Head et al. (1995, 1999) find that the regional presence of existing FDI is the most important agglomeration-related variable, in contrast to other studies that find a greater importance of agglomerations of domestic manufacturing and/or services firms. The main explanation for this difference in findings is that Head et al. analyse the location pattern of Japanese FDI, whereas the other studies analyse location decisions of FDI from several origin countries.[2] Having

---

2   Japanese FDI firms place great importance on the presence of existing FDI of their own country (see Friedman, Gerlowski and Silberman 1992).

said so, it is clear that the available empirical evidence strongly indicates that agglomeration economies play an important role as FDI location factor in a variety of host economies, indicating that FDI firms are attracted to regions within host economies that contain agglomerations of economic activity.

3. FDI in Mexico during Trade Liberalisation

Mexico belongs to the select group of developing countries that receive substantial flows of international investment. For instance, in the period 2003-2005, Mexico received over 8 percent of total FDI flows towards developing countries. Within Latin America, Mexico received about 20 percent of total inward FDI during this period, preceded only by Brazil.[3] Furthermore, Mexico represents one of the countries that have changed their policies towards FDI most drastically during the last 20 years. Following several economic crises in the 1970s and early 1980s, the Mexican government implemented drastic and rigorous changes to the country's development strategy, substituting policies of economic liberalisation and trade promotion for import substitution and government intervention. State-owned companies were sold of at a rapid rate, import restrictions were either abolished or severely relaxed and structural programmes were initiated to promote exporting activities (Gallagher and Zharsky 2004, ten Kate 1992, Weis 1992, Hanson 1998a, Loser and Kalter 1992). Also, the laws on inward FDI were changed several times to facilitate and actively promote the level of FDI in the Mexican economy (Pacheco-Lopez 2005, Ramirez 2002, 2003, Jordaan 2008a). In 1989 for instance, the number of sectors in which FDI participation was either prohibited or restricted was lowered. Also, the government made it easier for FDI firms to operate in Mexico without any ownership restrictions. The maquiladora programme received important new impulses, which led to large FDI flows into the northern part of Mexico (Sklair 1993). The creation of the NAFTA agreement also fuelled increasing levels of inward FDI into the country, caused by the growing level of economic integration between the member countries (e.g. Blomström and Kokko 1997, Cole and Ensign 2005, Love and Hidalgo 2000) and further relaxations of the main laws on inward FDI (Ramirez 2002, 2003, Cuevas et al. 2005, Pacheco-Lopez 2005, Thomas and Grosse 2001, ECLAC 1999). As a result of all these changes, foreign investors are now able to participate in many economic sectors with little or no restrictions. Sectors and activities where there remains scope for further liberalisation of FDI participation include telecommunications, land transport, coastal shipping and airports (OECD 2007).

As a result of these policy changes, Mexico experienced a dramatic increase in the level of inward international investment during the period of trade liberalisation. Figure 3.1 shows the developments of the main components of international

---

3 Based on data taken from the World Investment Directory. Latin America and the Caribbean, available at http://www.unctad.org.

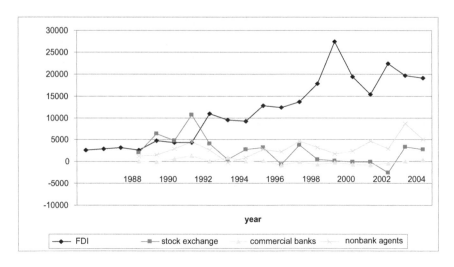

Figure 3.1   International investment into Mexico: 1985-2004

Source: Own calculations, based on data provided by Banxico.

investment into Mexico from the late 1980s to 2004. The figure reveals two important trends. First, it is clear that inward international investment increased rapidly after 1994, indicating the favourable effect of the creation of the NAFTA agreement on the level of international investment into Mexico. Second, whereas there was little difference in the relative importance of the various components of inward international investment prior to 1994, after the creation of NAFTA the relative importance of inward FDI has grown substantially, indicating the growing importance of FDI as main type of inward international investment into the Mexican economy. To further indicate the growing importance of this type of inward investment, the value of inward FDI stock as percentage of total GDP of the Mexican economy increased from 8.5 percent in 1990 to more than 27 percent in 2006 (see UNCTAD 2007).

Four additional remarks need to be made here. First, it is of course very difficult to identify the unique contribution of the creation of NAFTA on the marked increase in inward FDI. Undoubtedly, measures of FDI promotion and changes in the main investment laws during the late 1980s and early 1990s are partly behind the increase in inward FDI during the second half of the 1990s. Having said so, there are indications that the NAFTA effect has been substantial. For instance, Cuevas et al. (2005) estimate that the creation of the NAFTA agreement generated FDI flows into Mexico some 60 percent more than would have been the case if the trade agreement had not been established. Waldkirch (2003) conducts a related empirical study and concludes that the NAFTA agreement can be linked to a 40 percent increase in inward FDI. Second, FDI consists of new equity investment, reinvested earnings and inter-company loans. Overall, the first component has been the most important: during the period 1994-2006, new equity investment constituted more than 60 percent of total FDI.[4] The large share of new equity investment indicates that the majority of inward FDI led to the creation of new production facilities. Third, the US has traditionally been the most important source country of FDI. In 1980 for example, the share of US FDI in total inward FDI in Mexico amounted to more than 75 percent (Graham and Wada 2000). Although there has been an increase in participation in the Mexican economy by other countries in more recent years, the importance of the US as main source country has persisted. Statistics provided by the Secretaria de Economía show that over the period 1994-2005 the average share of the US in FDI flows into Mexico was well over 60 percent. Finally, the majority of FDI has taken place in the manufacturing sector. During the period 1994-2006, about 60 percent of inward FDI went into manufacturing industries, followed by 27 percent for financial and other services and 15 percent for commerce and transport and communication.

Within the manufacturing sector, there is substantial variation over the different activities. As Table 3.1 shows, four subsectors in particular received the lion share of FDI into the manufacturing sector during the period 1994-2006. The subsector with the largest share is machinery and equipment, which contains typical MNE activities such as the assembly and production of cars and car parts, televisions, personal computers and other electronic products. The subsector of food, beverages and tobacco received the second largest share, followed by chemical industries and the subsector of basic metals. One remark that needs to be made regarding the shares of the various subsectors in inward FDI flows is that they present a somewhat distorted picture of the actual distribution of FDI over the various activities, caused by differences in capital intensity of investments between these activities. In particular,

Table 3.1   FDI flows into manufacturing industries

Manufacturing Industries	1994-2001 (%)	2001-2006 (%)
Machinery and equipment	47.9	34
Food, beverages and tobacco	18.9	29
Chemical industries	12.9	17.9
Miscellaneous industries	7.8	0.6
Basic metals	5.1	9
Textiles and leather	3.8	2.2
Paper and paper products	2	2.5
Non-metallic products	1.3	3.8
Wood and wood products	0.4	1

Source: Own calculations, based on data provided by Secretaria de Economía.

---

4   Own calculations, based on additional data provided by the Secretaría de Economía and Banxico.

the reported share of the sector of textiles and leather is likely to be affected by this. In the period 1994-2001, 15 percent of all new foreign-owned firms were registered in this sector (Pacheco-Lopez 2005). The participation of this sector in the value of total inward FDI flows amounted to less than 4 percent during this period, caused by the fact that the average level of capital investment to create a firm in this sector is much lower compared to sectors such as chemical industries and basic metals. This means that Table 3.1 understates the importance of FDI in those sectors that use labour intensive production technologies.

FDI and the Maquiladora Programme

A central feature of foreign participation in the Mexican economy is that many foreign-owned firms participate in the maquiladora programme. In 1965, the Mexican government initiated the Border Industrialisation Program, attracting FDI firms to stimulate the generation of much needed manufacturing jobs in the northern states (Sklair 1993, CEPAL 1996, Wilson 1992). The special benefits that maquiladora firms obtain from participating in this programme is that they can import inputs from and re-export finished assembled products to the US tax-free; taxes are only paid on the value added that is generated in the production process located in Mexico. This programme has been attractive in particular to those MNEs whose production processes contain labour intensive stages, which can be located in Mexico to benefit from the difference in wage costs between the US and Mexico (Sklair 1993, South 1990, Wilson 1992).

Initially, the programme placed many restrictions on participating firms, including locational restrictions, ownership restrictions and strict regulations that all finished products had to be re-exported to the US. Over the years, in line with the ongoing liberalisation of the Mexican economy, these regulations have loosened substantially (South 1990). Currently, there are virtually no locational restrictions, participating firms are allowed to sell a substantial part of their production on the Mexican market and ownership conditions have become less strict. As a result, after the initial large dominance of US maquiladora firms, recent years have witnessed a growing number of so-called third country maquiladora firms, which are owned by MNEs from countries other than the US. Also, a substantial number of maquiladora firms are currently in operation under partial or even full Mexican ownership.

To indicate the overall importance of the maquiladora programme for the Mexican manufacturing sector, Table 3.2 presents the number of maquiladora firms, total maquiladora employment and the relative importance of maquiladora employment with respect to total manufacturing employment. As can be seen, the initial contribution of the maquiladora programme was modest. Following the start-up phase, the total number of jobs generated by maquiladora firms represented some 5.5 percent of total manufacturing employment. During the 1980s, the maquiladora programme entered a second stage, growing rapidly in both number of firms and number of created jobs. In 1988, 14 percent of Mexico's manufacturing labour

Table 3.2   Growing importance of maquiladora programme

Year	Number of maquiladora firms	Number of employees working for maquiladora firms	% Share in total manufacturing employment
1965	12	3,000	n.a.
1970	120	20,327	n.a.
1975	454	67,214	n.a.
1980	620	119,546	5.5
1988	1,396	369,489	14
1993	2,050	580,000	17.8
1998	2,590	1,014,006	21
2003	2,860	1,062,105	23.3
2006	2,810	1,202,134	25

Source: 1965-1988 South (1990), 1993 Cooney (2001), 1988-2006 own calculations based on data provided by INEGI and Economic Census, various years.

force was employed by firms participating in the maquiladora programme. At the end of the 1990s, this percentage had increased to 21 percent. The latest available statistics reinforce the impression that the maquiladora programme has become a central element of the Mexican manufacturing sector, as one out of every four manufacturing employees in the country is employed by a maquiladora firm.[5]

The Regional Distribution of FDI

The regional impact of the introduction of economic liberalisation and trade promotion in the 1980s has been distinctly heterogeneous. During the period of import substitution, a large agglomeration of economic activity developed in and around Mexico City. Following the introduction of trade liberalisation, the late 1980s and 1990s witnessed a substantial decrease in the share of this agglomeration in the Mexican economy. During the same period, the northern states sharing a border with the US benefited greatly from the opening up of the

---

5   Of course, the creation of the large number of jobs is only one, be it central, component of the impact of the maquiladora programme on the Mexican economy. The evaluation of other components of this impact is more mixed. Other important positive aspects are that the maquiladora programme has stimulated industrialisation processes in the northern states and that it has increased Mexico's participation on international export markets to a great extent, generating important revenues and also stimulating Mexican firms to become engaged in exporting activities. At the same time, maquiladora firms have a reputation to '...thrive upon very low wages, minimal labour standards, sell very little of their output in Mexico and buy no more than 3 percent of their materials, parts and components from Mexican suppliers' (Ramirez 2002: 870-871, see also Cypher 2001).

Mexican economy, experiencing substantial growth in several manufacturing industries (Hanson 1998a, Krugman and Elizondo 1996, Jordaan and Sánchez-Reaza 2006, Faber 2007). As a result of these changes, the regional distribution of the Mexican economy is now characterised by the existence of a limited number of regional production centres in the border states and in Mexico City.[6]

To see how the large influx of new FDI has been allocated over the 32 Mexican states during the last two decades, I collected data on the share of these states in total inward FDI flows, shown in Table 3.3. A central feature is the strong dominance of Mexico City as destination region of inward FDI. During the period 1989-2005, the average share of this region in total inward FDI ranged between 60 and 67 percent. In combination with the importance of Mexico City as agglomeration of manufacturing activity, the capital city is also the financial and political centre of the country, attracting a majority of FDI in finance and related services. Having said so, it is likely that the share of Mexico City in inward FDI is inflated. Although many foreign-owned firms have their headquarters in Mexico City to which investment flows are assigned, the new production sites and back offices that are generated by the investments may be located elsewhere in the country. Therefore, although there is no doubt that Mexico City is a very important destination region for FDI, it seems safe to assume that its relative importance is overstated in Table 3.3.

The second important feature of the regional distribution of FDI is that the border states have seen their share in inward FDI increase substantially. In the early years of trade liberalisation, this group of states received about 12 percent of total inward FDI flows. In more recent years, this share has more than doubled, reflecting the growing importance of these states as location for FDI firms. In fact, the importance of the border states is likely to be larger than indicated by their share in the value of inward FDI flows, as the border states incorporate many foreign-owned activities in industries that use labour intensive production technologies. Also, the border states are likely to incorporate foreign-owned activities that have their headquarters in Mexico City, to which investment flows are assigned. In comparison to the border states, the remaining states of Mexico have experienced a marked decrease in their participation in inward FDI flows. At the beginning of the period of trade liberalisation, these states received over 20 percent of inward FDI; the statistics for recent years show that this has dropped to 10 percent.

Next to the regional distribution of overall inward FDI, it is of course important to see where foreign-owned maquiladora firms locate. Table 3.4 presents the regional distribution of maquiladora employment for the period 1990-2004.

---

6   The diversified spatial outcome of the introduction of trade liberalisation is fully explored in Chapter 4. For the purpose of the present chapter, it is important to know that the main feature of these spatial changes has been the structural shift from a situation with one main agglomeration of economic activity in Mexico City to a situation where important shares of the Mexican economy are incorporated into a number of regional production centres in the north and centre of the country.

Table 3.3   FDI flows into Mexican states

State	1989-93 (%)	1994-99 (%)	2000-05 (%)
Mexico City	**67.3**	**58.5**	**65.5**
Distrito Federal	59.9	51.6	62
Estado de México	7.4	6.9	3.5
Border States	**12.2**	**30.6**	**24.2**
Baja California	3.0	7	2
Coahuila	0.7	1	1.3
Chihuahua	1.4	5.1	5.4
Nuevo León	3.9	12.6	12.1
Sonora	0.8	1.6	1.1
Tamaulipas	2.4	3.3	2.3
Other States	**20.5**	**10.9**	**10.3**
Aguascalientes	0.1	0.3	0.1
Baja California Sur	5.5	0.4	1
Campeche	0	0	0
Colima	0.8	0	0
Chiapas	0.1	0	0
Durango	1.6	0.2	0
Guanajuato	0.4	0.5	1.3
Guerrero	0.1	0.3	0
Hidalgo	0.1	0	0
Jalisco	3.8	2.4	2.2
Michoacán	0.3	0.2	0
Morelos	0.3	0.7	0
Nayarit	0	0	0.7
Oaxaca	0.1	0	0
Puebla	0.7	1.5	1.6
Queretaro	1.3	0.8	1.2
Quintana Roo	0.2	0.6	0.4
San Luis Potosi	1.1	1.1	0.8
Sinaloa	0.3	0.6	0
Tabasco	0	0.1	0
Tlaxcala	0.2	0.1	0
Veracruz	1.7	0	0.5
Yucatan	0.6	0.1	0.2
Zacatecas	0	0	0

Note: A score of 0 indicates that the share in inward FDI is negligible.

Source: Own calculations, based on data provided by Secretaria de Economía.

In contrast to its central role as indicated in Table 3.3, Mexico City plays no role as location for maquiladora firms. Initially, Mexico City was one of the regions from which maquiladora firms were prevented from locating. However, the lifting

Table 3.4   Regional employment maquiladora FDI (%)

Region	1990	1994	1998	2001	2004
Mexico City	**0.50**	**0.70**	**1.1**	**1.2**	**0.5**
Distrito Federal	0.13	0.13	0.14	0.2	0.1
Estado de México	0.37	0.57	0.96	1.0	0.4
Border States	**93.6**	**89.8**	**84.2**	**82.6**	**83.1**
Baja California	19.5	21.1	21.3	21.8	21.0
Coahuila	6.9	9.0	9.2	9.1	9.6
Chihuahua	37	29.1	25.9	24.1	24.5
Nuevo León	3.2	4.1	4.5	5.0	5.3
Sonora	8.6	8.5	8.8	8.2	7.2
Tamaulipas	18.4	18	14.5	14.4	15.5
Other States	**5.9**	**9.5**	**14.7**	**16.2**	**16.4**

Source: Own calculations, based on data in Sistema de Cuentas Nacionales de México. La Producción, Salarios, Empleo y Productividad de la Industria Maquiladora de Exportación. Aguascalientes: INEGI (various years).

of this location restriction in the 1990s and the possibility to sell a substantial share of production on the Mexican market have not lead to any increase in the presence of maquiladora firms in this region. In strong contrast, it is clear that the vast majority of these firms are located in the border states. Initially, the border states were earmarked as location for this type of FDI activity. The lifting of this restriction has led to only a modest decrease in the share of the border states in total maquiladora employment. Of course, proximity to the US, which is the main source of inputs and main destination market for assembled products, is the main reason for the continued preference of maquiladora firms to locate in these states. Having said so, although the dominance of the border states has persisted, the third interesting feature of the information presented in Table 3.4 is that the other states have seen their share in maquiladora activity increase, from 6 percent of total maquiladora employment in 1990 to over 16 percent in 2004. Additional data published by INEGI in its annual publication on the maquiladora industries shows that in particular firms operating in the industries of textiles and leather are seeking out regions located further away from the US border (see also South 2006). States that have benefited most from this recent development are Jalisco, Yucatan, Durango and Puebla.

4. FDI Location in Mexico: Data, Methodology and Definition of Variables

The period of trade liberalisation is characterised by a situation where economic activity has become concentrated in a limited number of agglomerations in Mexico City and the border states. The large and growing levels of inward FDI have also shown a tendency to concentrate in these same states, suggesting that agglomeration economies are an important location factor for new FDI firms. Two recent studies on FDI in Mexico present initial empirical evidence that supports the notion that FDI firms prefer to locate in regions that contain agglomerations of economic activity. Mollick et al. (2006) estimate determinants of inward FDI flows for 22 states during the period 1994-2001 in a panel data setting, producing evidence that regional infrastructure and a proxy variable for agglomeration economies, in the form of the manufacturing share in state level production, both have a positive effect on the level of regional inward FDI. Jordaan (2008a) presents a more inclusive panel data study on FDI location in Mexico, covering regional FDI inflows into all 32 states for the period 1989-2006. Also, this study is more careful in estimating separate effects of agglomeration economies from regional manufacturing and from services, as well as the effect of the regional presence of existing FDI firms. The findings offer important evidence that agglomeration economies from manufacturing and from the regional presence of existing FDI firms both have a positive effect on the regional level of new inward FDI flows (see Jordaan 2008a).

An important disadvantage of using information on FDI flows to identify location factors of FDI firms is of course that the estimations are likely to be affected by aggregation bias and measurement errors. Also, the regional distribution of FDI flows may be distorted when a few new FDI projects involve very large amounts of capital investment. To obtain further quantitative evidence on which state characteristics are important for the location decision of FDI and what role agglomeration economies play, I obtained unpublished and thus far unexplored data from the Secretaria de Economía, who kept track of the location decisions of almost 3,500 new foreign-owned manufacturing firms during the period 1994-1999. The nature of the dataset makes that I can use the conditional logit model, as originally introduced by McFadden (1974), to obtain statistical evidence on which state characteristics affect the location choice of new FDI firms in Mexico. The process of FDI location can be understood as the outcome of a profit maximisation strategy, where a foreign-owned firm chooses a region in a host economy with the highest expected profit (Carlton 1983, Bartik 1985). The regional distribution of new FDI firms is seen as the outcome of a set of discrete choices by these firms between the regions in a host economy. The probability that a FDI firm chooses state k in Mexico can be depicted as:

(1) $\quad \text{Prob}_k = \dfrac{e^{Xk\beta}}{\displaystyle\sum_{\lambda=1}^{32} e^{X\lambda\beta}}$;

where X contains state characteristics that are hypothesised to influence the firm's choice between the 32 Mexican states. As discussed earlier, four types of regional characteristics tend to be included in studies on FDI location: regional demand, regional cost factors, regional policies designed to facilitate or attract foreign investment and variables capturing the presence of agglomeration economies. I focus on the effects of regional demand, production costs and agglomeration economies, as quantitative data on FDI policies in Mexican regions is unavailable. In addition to these location factors, I also estimate the effect of regional distance to Mexico City and to the US, as they represent the main destination markets for most manufacturing firms (Hanson 1998a, Jordaan and Sanchez-Reaza 2006, Faber 2007).

Independent Variables

Table 3.5 presents the regional characteristics that I use in the empirical analysis. The first factor that needs to be included is market demand (e.g. Coughlin 1991, Head et al. 1999, Coughlin and Segev 2000). A new foreign-owned firm is more likely to choose a region when it expects this region to have a high demand for its products. I measure regional demand as the value of state level GDP.

The second factor is the regional level of production costs. In line with the majority of related studies, I focus on production costs associated with the production factor labour, captured by the wage level (e.g. Friedman et al. 1996, Coughlin and Segev 2000, Crozet et al. 2004, Hilber and Voicu 2009). The hypothesis is that foreign firms are attracted to regions with a relative low wage level. I measure the regional wage level as the total wage bill of the regional manufacturing labour force divided by the total number of regional manufacturing employees. I also need to control for the feature that the wage level incorporates both production costs and productivity. To do so, I use two variables that capture regional differences in human capital. One variable captures state-wide human capital, measured as the average level of schooling attainment of the regional economic active population. The second variable that I use captures the level of labour quality of manufacturing workers, measured as the total number of white collar employees divided by the total number of blue collar employees (see Jordaan 2005).

The third type of regional characteristic that I include in the regression model is the regional presence of agglomeration economies. The data allow me to make a detailed estimation of the effect of agglomeration economies from several sources. In particular, I distinguish between agglomerations of manufacturing firms, wholesale and distributors and financial services. Furthermore, I distinguish between Mexican-owned and foreign-owned agglomerations for these separate sources of agglomeration economies. In line with previous research, I calculate the agglomeration variables as the regional number of manufacturing firms, wholesale and distributors and financial firms. In calculating the number of firms for these different activities for Mexican firms, I omit firms that have fewer than 20 employees. The reason for doing so is that the Mexican economy is characterised by a considerable level of duality (Blomström 1989), where traditional and modern segments co-exist within regional industries. The traditional segment is characterised by a predominance of small and micro-sized firms, whereas the

Table 3.5   State characteristics

Variable	Definition	
Market demand	State gross domestic product	(1)
Wages	$\dfrac{(wages\,manufacturing\ sector)}{(number\,of\,manufacturing\,employees)}$	(2)
Schooling	Number of years schooling regional economic active population	(1)
Labour Quality	$\dfrac{(number\,of\,white\,collar\,manufacturing\,employees)}{(number\,of\,blue\,collar\,menufacturing\,employees)}$	(2)
Agglomeration Mexican firms – manufacturing	Number of Mexican-owned manufacturing firms (more than 20 employees)	(3)
Agglomeration Mexican firms – financial services	Number of Mexican-owned financial services firms (more than 20 employees)	(3)
Agglomeration Mexican firms – wholesale and distributors	Number of Mexican-owned distributors and wholesale commercial firms (more than 20 employees)	(3)
Agglomeration foreign-owned firms – manufacturing	Number of foreign-owned manufacturing firms	(3)
Agglomeration foreign-owned firms – financial services	Number of foreign-owned financial services firms	(3)
Agglomeration foreign-owned firms – wholesale and distributors	Number of foreign-owned distributors and wholesale commercial firms	(3)
Distance to Mexico City	Distance in kilometres between state capital city and Mexico City	(4)
Distance to US	Distance in kilometres between state capital city and nearest border crossing with US	(4)

Note: Schooling is for 1990; all other variables are for 1993. All variables are in logs.

Sources: (1) www.inegi.gob.mx; (2) 1994 Economic Census; (3) Unpublished data, provided by INEGI; (4) Calculated with data from http://www.trace-sc.com/maps_en.htm.

modern segment contains mostly medium-sized and large firms. As it is likely that FDI firms are only influenced in their location choice by the cross-regional variation of the modern segment, I count only those Mexican firms that employ at least 20 employees.[7]

---

7   One could argue that the regional size of the traditional segment may deter new FDI firms. In preliminary regressions, I did experiment with a variable capturing the cross-regional variation of micro-sized firms, but this variable does not carry a significant coefficient.

Finally, I include two distance variables in the regression model, in the form of regional distance to Mexico City and the US border. During the period of trade liberalisation, Mexico City and the US have become the two main destination markets for manufacturing firms in the country. This means that FDI firms may perceive regional proximity to one of these markets as an important location factor. I measure these variables as the distance in kilometres between the state capital cities and Mexico City or the nearest US border crossing. For this, I use the 'as the crow flies' assumption, as there is no information on travel times.

5. Empirical Findings

Table 3.6 presents the main findings from estimating a variety of specifications of the econometrical model on FDI location decisions, using maximum likelihood techniques. The first column with results contains the findings from estimating the model containing regional demand and the labour-related variables. Most of the estimated associations are in line with expectations. The level of regional demand has a positive effect on the probability that a region is selected by new FDI. The estimated effect of regional wages is negative, indicating that the level of labour costs has a negative effect on the arrival of new FDI firms. Furthermore, the estimated effect of schooling is positive, indicating the positive effect of regional human capital. The only variable with an estimated effect contrary to expectations is labour quality, which carries a negative coefficient.[8]

The second column with results contains the findings from estimating the model that includes the various agglomeration variables of Mexican firms. Although the magnitude of the estimated effect of the other variables changes somewhat, the nature of their effect stays the same. The exception to this is the estimated effect of regional labour quality, which turns positive. Looking at the coefficients of the three agglomeration variables, the findings indicate two different effects. On the one hand, the regional presence of agglomerations of Mexican manufacturing firms and professional input providers increase the probability that a state is selected by a new FDI firm. On the other hand, a large regional presence of financial services lowers this probability. This finding constitutes an important qualification to the findings on FDI location factors obtained from estimating determinants of the regional distribution of FDI flows (see Jordaan 2008a). In that study, the value of regional FDI flows is positively associated with the presence of an agglomeration of manufacturing activity. The present findings show that in fact the effect of agglomeration economies is more nuanced: the regional presence

---

8   Following Hilber and Voicu (2009), I also experimented with variables that capture labour relations, in the form of the regional number of official strikes and the regional number of other registered labour conflicts. The estimated effect of these variables is insignificant. Also, following Mollick et al. (2006) and Jordaan (2008a), I estimated the effect of several regional infrastructure variables; again, the estimated effect of these variables fails to reach significance.

Table 3.6 Empirical findings on FDI location factors

	1	2	3	4	5	6	7	8
	Full sample				Omitting Mexico City			Nested logit
GDP	1.10 (0.04)a	0.52 (0.16)a	1.30 (0.23)a	0.65 (0.25)a	-1.83 (0.20)a	-1.02 (0.20)a	-0.17 (0.25)	0.39 (0.29)
Wages	-0.21 (0.10)b	-2.24 (0.17)a	-1.71 (0.18)a	-0.83 (0.17)a	-0.49 (0.18)a	-0.50 (0.17)a	-0.74 (0.18)a	-0.93 (0.20)a
Schooling	0.48 (0.02)a	0.57 (0.03)a	0.39 (0.03)a	0.25 (0.03)a	0.18 (0.02)a	0.25 (0.03)a	0.11 (0.03)a	0.28 (0.03)a
Labour quality	-1.23 (0.11)a	0.98 (0.18)a	0.78 (0.22)a	0.77 (0.25)a	-0.24 (0.17)	1.42 (0.20)a	1.58 (0.20)a	0.71 (0.28)a
Agglomeration Mexican manufacturing	—	1.22 (0.08)a	1.55 (0.18)a	1.48 (0.17)a	1.72 (0.08)a	0.84 (0.11)a	1.22 (0.12)a	1.03 (0.21)a
Agglomeration Mexican financial services	—	-0.65 (0.16)a	-1.32 (0.17)a	-0.61 (0.17)a	-0.09 (0.009)a	0.004 (0.01)	-0.67 (0.17)a	-0.78 (0.17)a
Agglomeration Mexican distributors	—	0.07 (0.01)a	0.06 (0.01)a	0.06 (0.01)a	0.45 (0.18)a	-0.09 (0.16)	0.06 (0.01)a	0.05 (0.01)a
Agglomeration foreign-owned manufacturing	—	—	0.47 (0.07)a	0.17 (0.08)b		0.37 (0.06)a	0.30 (0.11)a	0.06 (0.10)
Agglomeration foreign-owned financial services	—	—	-0.04 (0.008)a	-0.005 (0.0008)a		-0.11 (0.007)a	-0.02 (0.004)a	-0.01 (0.009)a

Table 3.6 continued   Empirical findings on FDI location factors

	1	2	3	4	5	6	7	8
	Full sample				Omitting Mexico City			Nested logit
Agglomeration foreign-owned distributors	–	–	0.79 (0.10)a	0.35 (0.10)a	–	0.90 (0.08)a	0.05 (0.01)a	0.50 (0.12)a
Distance to Mexico City	–	–	–	-0.03 (0.003)a	–	–	0.15 (0.03)a	-0.02 (0.004)a
Distance to US	–	–	–	-0.02 (0.003)a	–	–	-0.02 (0.004)a	-0.03 (0.0038)a
Inclusive value Nested logit	–	–	–	–	–	–	–	0.42 (0.05)a
Log likelihood	-7852.8	-7783.7	-7593.4	-7587.2	-5472.5	-5252.3	-5164.1	-7543.16
Number of choices	32	32	32	32	30	30	30	2, 30
Number of investors	3492	3492	3492	3492	1928	1928	1928	3492

Notes: Standard errors in parentheses. a, b and c indicate significance levels at 1, 5 and 10%. Nested logit estimation: firms choose first between Mexico City and the rest of the country, after which they select a region within one of these two broad regions. Inclusive value of nested logit estimation concerns this tree structure of Mexico City versus the rest of the country.

of financial services deters new manufacturing FDI, whereas agglomerations of manufacturing firms and professional input providers both attract new FDI firms. In combination, these findings suggest that foreign-owned manufacturing firms are attracted in particular to regions that are specialised in manufacturing and supporting commercial input provision.

The third column with results contains the results from adding the FDI-based agglomeration variables to the regression model. Similar to the effect of the agglomeration variables of Mexican firms, the regional presence of foreign-owned manufacturing firms and foreign-owned commercial input providers positively influence the probability that a state is selected by a new FDI firm. The regional presence of foreign-owned financial services firms lowers this probability. Previous studies that have estimated the effect from existing regional FDI have all concentrated on the effect of the regional presence of manufacturing FDI, presenting evidence that this regional presence has a positive effect on the arrival of new FDI firms. Not only do the present findings confirm this positive effect, they also indicate that a similar positive effect is associated with the regional presence of foreign-owned input suppliers, whereas the regional presence of foreign-owned financial services has a negative effect on the arrival of new FDI firms.

To understand the negative effect from the regional presence of financial services, it is important to consider that an agglomeration is a costly place to operate in. The presence of agglomeration economies serves partly to compensate for the higher costs of e.g. labour and land that result from the high level of geographical concentration of economic activity (see Henderson 1988). This suggests that a likely explanation for the estimated negative effect of regional financial services is that foreign-owned manufacturing firms avoid locating in regions with a large presence of financial services, as the advantages of this type of agglomeration are likely to be less relevant to them, whilst they would still have to pay the premium for locating in such an agglomeration of activity. The reason why financial services may be less relevant to foreign-owned firms is that they obtain these services from within the structure of the MNEs to which they belong.

Finally, column four contains the results from estimating the full econometrical model, including the two variables that capture regional distance to Mexico City and the US. Both these distance variables carry negatively signed significant coefficients, in line with expectations. This suggest that foreign-owned firms do perceive proximity to their main destination market as important independent location factor, even after controlling for the effects of the other state characteristics. The nature of the estimated effect of these other control variables does not change. Looking at the relative importance of the various agglomeration variables that have a positive effect on the probability that a region is selected by a new FDI firm, the regional presence of domestic manufacturing firms has the largest estimated positive effect, a finding which is in line with findings presented by e.g. Crozet et al. (2004) and Guimarães et al. (2000). The presence of foreign-owned material input providers and foreign-owned manufacturing firms have the second largest effect, followed by the effect of the regional presence of domestically-owned input providers.

Robustness Checks

A central issue with estimating conditional logit models is that it rests on the assumption that alternatives are equal substitutes. In the present analysis, this means that all regions are equal substitutes for a new foreign-owned firm. This assumption follows from the requirement of the conditional logit model of independence of irrelevant alternatives, which implies that the relative probability of choosing between two alternatives should not depend on the availability or characteristics of other alternatives. A violation of this property will generate biased estimates (see Head et al. 1995).

There are two reasons why this assumption of equal alternatives may not hold. First, one could argue that Mexico City constitutes a region that is structurally different from all other Mexican regions, as it is the political and financial centre and also constitutes the main destination market in the country. Second, the sample of new FDI plants contains two types of FDI firm: firms that do and firms that do not participate in the maquiladora programme. As it seems fair to assume that there will be important differences between these two types of FDI firm, it is also likely that both types of firm will attach different importance to location factors.

One way to address both these issues is to estimate the econometrical model on a restricted sample that omits all FDI firms that locate in Mexico City. Doing so eliminates a possible 'Mexico City effect' and focuses on those FDI firms that are most likely to be export-oriented, as firms that produce for the Mexican market are more likely to locate in and around Mexico City (see Faber 2007). The results of estimating the regression model on the restricted sample are presented in columns five through seven. Comparing the findings for the restricted sample with the findings for the full sample, there are some important differences and similarities. Most importantly, the estimated effect of regional demand is either negative or insignificant, indicating that regional demand is not an important location factor for those FDI firms that are producing for the international market. Also, schooling appears to be less important for this type of FDI, whereas the importance of the quality of the manufacturing labour force is more important.

Next, looking at the effect of the various agglomeration variables, the regional presence of Mexican and foreign-owned manufacturing firms and material input providers all persist to have a significant positive effect on the probability that a region is selected by new FDI. Again, the regional presence of an agglomeration of financial services lowers this probability. This last finding is particularly interesting, when considering that this estimated negative effect is maintained for the sample that omits the financial centre Mexico City. This suggests that the estimated negative effect from the regional presence of financial services for the full sample is not caused solely by a 'Mexico City effect'. Finally, the estimated negative effect of regional distance to the US and the estimated positive effect of regional distance to Mexico City indicates that export oriented FDI firms prefer to locate in proximity to the international market.

An alternative way to address the possible violation of the assumption of independence of irrelevant alternatives is to estimate nested logit models. The assumption that all regions represent equal alternatives does not hold when there is a hierarchy in the decision making process of FDI firms. In particular, foreign-owned firms may decide first whether or not to locate in Mexico City. If they choose to locate elsewhere, they then choose among the remaining 30 other states. The estimation of a nested logit model controls for the presence of such a hierarchy in decision making.

The findings from the nested logit model are presented in column eight. The inclusive value of the nested logit model is significant, indicating that the assumed hierarchy in the decision making process is accepted. The findings show that the effect of the level of regional demand is insignificant. This indicates that new FDI plants are not influenced by this location factor, once they have decided between locating in Mexico City and the rest of the country. The explanation for this difference in finding between the standard and the nested logit model estimation is the very large difference between the level of regional demand of Mexico City and the other Mexican states. The level of GDP in Mexico City is almost seven times larger than average state GDP in the rest of the country, suggesting that the initial estimated positive effect of regional demand may have been caused by this large regional difference in market demand.[9] In contrast to this difference in findings between the standard and nested logit model specifications, the other non-agglomeration variables carry significant coefficients with signs similar to the previous estimations. Looking at the importance of the various agglomeration variables, the nature of their estimated effect remains stable when estimating the nested logit model specification. Also, the relative importance of most of the agglomeration variables is not subject to significant change, compared to the original findings for the full model as presented in column four. Therefore, although the findings indicate that the tree structure in the location decision making process of new manufacturing firms is important, the findings persist to indicate that the regional presence of Mexican- and foreign-owned manufacturing firms and material input providers have a positive effect on the probability that a region is selected by new FDI and that the regional presence of financial services lowers this probability.

Inter-regional Effects of Demand and Agglomeration

One potential omission of the analysis presented thus far is that the econometrical model does not allow for the possibility that the effects of regional demand and of agglomeration economies have a spatial dimension. Considering for instance the location factor of regional demand, the inclusion of inter-regional demand

---

9   Crozet et al. (2004) provide a similar explanation of their findings from nested logit models on FDI location factors in France, concerning the dichotomy between Paris and other French regions.

flows produces a variable capturing the full market potential of a region as follows (Harris 1954, also Head and Mayer 2004):

$$\text{MarketPotential}_k = \text{GDP}_k + \sum_{k'=1;k'\neq k}^{k} W\ \text{GDP}_k;$$

where W is a distance matrix containing spatial weights w_{ik}, capturing the relation between geographical space and inter-regional demand flows between regions i and k (see Anselin 1988). The use of this spatially adjusted regional demand variable controls for the possibility that there are multi-region areas within Mexico that have an aggregate demand level that affects the location decision of new FDI firms. In a similar fashion, the effect of agglomeration economies may also have a multi-regional dimension. Economic activity may be agglomerated in a multi-regional area within Mexico, or agglomeration economies generated in one region may transcend spatially to other regions. To assess whether such spatial agglomeration economies influence the location process of new FDI firms, I use the same calculation as for MarketPotential:

$$\text{Agglomeration}_k = \text{Agglomeration}_k + \sum_{k'=1;k'\neq k}^{k} W\ \text{Agglomeration}_k;$$

where agglomeration stands for the variables that indicate the regional presence of agglomerations of manufacturing, input provision and financial services, separated again between Mexican and FDI firms.

Important to consider is that it is not possible to determine *a priori* what the exact relation is between spatial flows of demand or agglomeration economies and geographical space. Therefore, empirical estimations of such spatial effects should assess the effect of several specifications of the distance-based philtre mechanism, as captured by the spatial weights w_{ik} (Anselin 1988, Bode 2004, Jordaan 2008b). In the present estimations, I estimate the spatially augmented regression model with three alternative spatial decay specifications. The first specification is the first order contiguity assumption, which holds that spatial effects may only occur between states that share a border. In terms of the variable of market potential for instance, this means that, for a given region, regional demand in those states that it shares a border with is assumed to possibly affect the probability that the region is selected by new FDI firms. This specification means that the spatial weights are given the value 1 for those states that share a border with the state, and 0 otherwise. The second specification that I use is the second order contiguity assumption. This assumption entails that, for a given state, states that share a border with the state are included, as well as those states that share a border with the states that are included under the first order contiguity assumption. Both the first and second order contiguity specification rest on the assumption that there is a geographical cut off point in the spatial transmission of demand and agglomeration effects; the difference between the first and second order contiguity assumption is that the latter assumption allows for the existence of geographically more extended spatial effects. In contrast to the first and second order contiguity assumptions, the third specification that I use is to specify the spatial weights w_{ik} as the inverse of distance between the states. This assumption envisages a continuous negative relation between geographical space and spatial demand and agglomeration effects. Following Crozet et al. (2004) and Head and Mayer (2004), I specify this gravity-like specification as the inverse of distance between the states measured in number of kilometres between the state capital cities.

Table 3.7 presents the findings for the demand and agglomeration economies variables from estimating the regression model that incorporates inter-regional demand and agglomeration effects. Looking first at the results from using the contiguity_1 assumption, the findings reflect three important features. First, the estimated effect of the market potential variable is insignificant, indicating that there are no inter-regional demand flows that affect FDI location decisions. Second, the spatially-adjusted agglomeration variables all carry coefficients that are significantly larger than those obtained from the non-spatially adjusted regression models. This significant increase in the magnitude of the estimated effect of the agglomeration variables suggests that there is a spatial dimension to the effect of agglomeration economies on the location decision of FDI firms. Third, the feature that spatial agglomeration economies are identified using the first order contiguity assumption indicates that geographical proximity plays an important role for these spatial effects.

Perhaps in line with this latter point, the findings from using the second order contiguity assumption are much less satisfactory. Although the nested logit specification persists to be significant, there is a considerable drop in the inclusive value parameter. Also, the estimated effect of agglomeration economies from Mexican manufacturing firms has decreased considerably, the estimated effects of agglomeration of foreign-owned manufacturing firms and distributors have become insignificant and the estimated effect of the regional presence of foreign-owned financial firms carries the wrong sign. In light of these poor results, the conclusion seems warranted that the second order contiguity assumption does not work very well in identifying the model.

In contrast to this, the findings from using inter-regional distances as spatial philtre mechanism are more satisfactory. Again, the findings suggest that there is no spatial demand effect. The significant increase in the magnitude of the coefficients of most of the agglomeration variables indicates that spatial dimensions of agglomeration economies do play a role in the location process of FDI firms. Also, there are some changes in the relative importance of the various agglomeration variables. In particular, the regional presence of Mexican manufacturing firms and foreign-owned suppliers of material inputs now carry estimated positive coefficients of a similar magnitude. This change in relative importance of these variables does not alter the fact that agglomeration economies from Mexican and foreign-owned

Table 3.7   Spatial effects of demand and agglomeration economies on FDI location behaviour

Variables	No spatial effects	Cont_1	Cont_2	Inter-regional distance
Regional demand	0.39 (0.29)	0.009 (0.19)	0.77 (0.26)a	0.11 (0.27)
Agglomeration Mexican manufacturing	1.03 (0.21)a	3.57 ** (0.32)a	0.14 * (0.13)	1.57 ** (0.21)a
Agglomeration Mexican financial services	-0.78 (0.17)a	-2.80 ** (0.28)a	-3.98 ** (0.78)a	-0.29 (0.18)
Agglomeration Mexican distributors	0.05 (0.01)a	0.52 ** (0.13)a	1.71 ** (0.40)a	0.52 ** (0.06)a
Agglomeration foreign-owned manufacturing	0.06 (0.10)	1.55 ** (0.15)a	-0.34 (0.22)	0.47 ** (0.14)a
Agglomeration foreign-owned financial services	-0.01 (0.009)a	-1.53 ** (0.34)a	1.99 ** (1.02)b	-1.32 ** (0.14)a
Agglomeration foreign-owned distributors	0.50 (0.12)a	1.32 ** (0.15)a	0.12 (0.09)	1.57 ** (0.21)a
Inclusive value	0.42 (0.05)a	0.58 (0.02)a	0.13 (0.02)a	0.62 (0.03)a
Log likelihood	-7543.16	-7457.79	-7534.11	-7473.66
Number of choices	2, 30	2, 30	2, 30	2, 30
Number of investors	3492	3492	3492	3492

Notes: Standard errors in parentheses. a, b and c indicate significance levels at 1, 5 and 10%.

Cont_1 refers to distance matrix based on first order contiguity; Cont_2 refers to distance matrix based on second order contiguity; inter-regional distance refers to distance matrix based on inverse of distance in kilometres between state capital cities.

Column labelled No spatial effects is a replication of column 8, Table 3.6.; Nested logit specification of estimations of Table 3.7 is similar to Table 3.6, column 8.

* Indicates that coefficient is significantly smaller than estimation without spatial effects.

** Indicates that coefficient is significantly larger than estimation without spatial effects.

The use of the nested logit specification is supported by the significance of the inclusive value parameter in all four estimations. (I also estimated the effect of the spatial variables using the standard conditional logit specification. The findings do not differ substantially from those presented in Table 3.7.)

The estimated regression models also contain regional demand, wages, schooling, labour quality and regional distance to Mexico City and the US.

Distance matrices are row standardised.

manufacturing firms and input suppliers all continue to have a positive effect on the probability that a region is selected by new FDI firms, whereas the regional presence of financial services lowers this probability. The statistically identified importance of spatial agglomeration economies with the spatial decay effect based on inter-regional distances indicates that spatial externalities are negatively related to geographical distance in a continuous fashion. This suggests that, in addition to the finding that geographical proximity is important for spatial agglomeration effects, there is also a more general relation between agglomeration effects and geographical space, where spatial agglomeration economies decrease with increasing inter-regional distance.

6. Conclusions

Mexico belongs to the select group of countries that receive a majority of FDI flows towards developing countries. Furthermore, following the introduction of trade promotion and economic liberalisation in the late 1980s, the level of inward FDI into the Mexican economy has increased sharply. During the same period, there have been important spatial changes within the Mexican economy. With the relative demise of Mexico City and the rapid development of states that share a border with the US, the regional distribution of economic activity in Mexico now consists of a limited number of regional production centres in the north and the centre of the country. Importantly, the large influx of new FDI firms has shown a trend to gravitate towards these regional production centres, suggesting that FDI firms are attracted to agglomerations of economic activity within Mexico.

In order to establish whether the regional presence of an agglomeration of economic activity does indeed play a role as location factor for foreign-owned firms, I use a new dataset, containing the location decisions of a large set of new manufacturing firms during the second half of the 1990s, to conduct an empirical analysis of FDI location factors. In the analysis, I have taken great care to distinguish between agglomeration economies from manufacturing, material input provision and financial services, and I calculate these agglomeration variables for Mexican and foreign-owned firms separately.

Based on the estimation of conditional logit models, the analysis produces two important sets of empirical findings. First, looking at the findings for the full sample of new manufacturing firms, the evidence suggests that state characteristics such as demand, wages, schooling, labour quality and regional distance to main destination markets all influence the location decision of new FDI firms. Furthermore, the findings also indicate that agglomeration economies play an important role. There are two main effects from the set of agglomeration economies variables. On the one hand, the regional presence of agglomerations of manufacturing firms and suppliers of material inputs both have a positive effect on the probability that a region is selected by new FDI firms. Importantly, the estimations identify separate positive effects of agglomerations of Mexican and FDI firms. On the other hand, the regional presence of an agglomeration of financial services firms lowers the probability that a region is selected by new foreign-owned manufacturing firms. In combination, these findings indicate that foreign-owned firms have gravitated towards regions that are specialised in manufacturing and supporting input provision. Importantly, the findings on the importance of the various types of agglomeration economies persist when I estimate the econometrical model for those foreign-owned firms that are most likely to be producing for the international market. The only variable that does not carry a significant coefficient in the estimations for this restricted sample is regional demand, which is in line with the export orientation of these firms. Furthermore, the estimated effect of the agglomeration variables does not change when I use a nested logit specification which controls for hierarchical decision making by FDI firms.

Second, the analysis addresses the possibility that effects from regional demand and from agglomeration economies have a multi-regional character. In particular, based on two different specifications of the relation between inter-regional flows of demand and agglomeration economies with geographical distance, the empirical findings indicate that agglomeration economies have spatial dimensions. Comparing the findings from the non-spatial with findings from the spatial regression model indicates that the magnitude of the estimated coefficients of the agglomeration variables is significantly larger with the latter type of regression model, indicating that spatial agglomeration economies play a role in the location decision of FDI firms.

Summing up, the empirical findings contain substantial indications that agglomeration economies are important for new FDI firms. FDI firms locate in agglomerations of economic activity within Mexico, and the statistical estimations of FDI location factors indicate that, controlling for the effect of several other regional characteristics, the probability that a region is selected by new FDI firms is significantly positively associated with the regional presence of an agglomeration of manufacturing activity and supporting input provision. Interestingly, the findings also contain evidence that agglomeration economies appear to have a multi-regional dimension. Overall, the findings from the empirical analysis of FDI location behaviour reflect that the phenomenon of agglomeration is a clear element of the business environment of FDI firms. Furthermore, these findings serve to underline the importance of the hypothesis that industry agglomeration or geographical proximity between FDI and Mexican firms may influence FDI externalities. Foreign-owned firms are concentrated in agglomerations of activity within Mexico, indicating the need for empirical verification of the relation between agglomeration and FDI spillovers.

Chapter 4

Trade Liberalisation and Regional Growth in Mexico: How Important are Agglomeration Economies and Foreign Direct Investment?

1. Introduction

Following several economic crises in the 1970s and early 1980s, the Mexican government substituted the development strategy of economic liberalisation and trade promotion for import substitution and government intervention. This drastic change in development strategy generated far-reaching changes in the Mexican economy, including a marked increase in the level of competition, a sharp decrease in the number of state-owned companies and a rapidly growing participation in international trade with the US and other countries. Furthermore, as discussed in the previous chapter, the change in development strategy also fostered a dramatic increase in the level of inward FDI into the Mexico economy.

A central feature of this period of economic liberalisation and trade promotion has been the distinct spatial dimensions of the changes in the Mexican economy. This feature was discussed briefly in the previous chapter, in relation to the location behaviour of new FDI firms. In this chapter, I discuss the spatial changes of the Mexican economy more in depth. Several studies have described the scale and nature of the changes in the location pattern of industries within Mexico. Also, there is substantial empirical evidence on the nature of regional growth regimes during the periods of import substitution and trade liberalisation and on the role of several regional characteristics as drivers of regional growth. However, the potentially important regional growth effects from FDI and agglomeration economies have received only limited attention. This is the more remarkable, given that growing levels of FDI and the changing nature of the regional distribution of economic activity have been two central features of the era of trade liberalisation. In light of this, the purpose of this chapter is two-fold. The first aim is to obtain a good impression of the spatial dimensions of the structural economic changes that have occurred in Mexico since the late 1980s. The second aim is to conduct a new empirical study on determinants of regional growth to identify the roles of the changing regional distribution of economic activity and of the growing presence of foreign-owned manufacturing firms.

The chapter is constructed as follows. In section 2 I present and discuss several indicators of the main spatial changes of the Mexican economy that occurred following the introduction of trade liberalisation. I also discuss in more detail the available evidence on determinants of changing location patterns of manufacturing activity and regional growth processes. Section 3 contains a description of the dataset that I built for the empirical analysis of this chapter. This dataset consists of state level observations for several time periods during the period 1988-2004. In addition to a variety of regional characteristics that have also been used in previous studies on regional growth in Mexico, the dataset contains several variables that capture agglomeration economies and the level and type of regional FDI. Section 4 presents the main empirical findings on the role of regional characteristics as drivers of regional growth, identified within the framework of conditional convergence growth regressions. The findings provide important new indications on which regional characteristics have been important factors for regional growth in Mexico during the last 20 years, including the effects from agglomeration and regional FDI. In addition to this, I also present findings on the spatial dimensions of agglomeration and FDI growth effects, containing important evidence on the nature and extent to which growth effects of these two elements of trade liberalisation are contained within regions or alternatively are also transmitted across geographical space. Finally, section 5 summarises and concludes.

2. Trade and the Location of Industries

The process of industrialisation of the Mexican economy was initiated during the 1940s. Prior to this, the Mexican economy was relatively open and characterised by a large dependence on the production of primary goods (Reynolds 1970). In the second half of the 1940s, in line with government policies of most other large Latin American countries, the Mexican government started formally with the implementation of a development strategy based on import substitution and government intervention. During the 1950s and 1960s, the development strategy of import substitution fostered a rapid development of the manufacturing sector. For instance, manufacturing output increased by an average annual growth rate of 7.5 percent during the period 1950-1981 (Bethell 1984). As a result, the share of the manufacturing sector in Mexico's total GDP increased steadily from less than 13 percent in 1930 to more than 23 percent in the 1970s (Hanson 1997). However, during the 1970s and early 1980s the Mexican economy was subject to several economic crises. In particular, following the boom in oil revenues in the 1970s, the Mexican economy was hit hard by the decrease in oil prices during the 1980s, culminating into a situation where Mexico became unable to make interest payments on its international loans (Cárdenas 1996).

In response to this situation, the Mexican government initiated a sudden and drastic change in its official development strategy (ten Kate 1992, Cárdenas 1996), replacing the strategy of import substitution by a strategy of economic liberalisation and trade promotion. This change in development strategy entailed a rapid decrease of restrictions on imports and exports (ten Kate 1992), a large decrease in the number of state-owned companies and the facilitation and active promotion of increasing levels of international investment into the country, including increased efforts to promote the participation by FDI firms in the maquiladora programme (Loser and Kalter 1992, Pacheco-Lopez 2005, Ramirez 2002, 2003, Jordaan 2008a). In 1988, the first stage of economic liberalisation was consolidated when Mexico joined the General Agreement of Tariffs and Trade (GATT). The second stage was completed in 1994, with the creation of NAFTA by the US, Canada and Mexico. Following this, growing levels of economic integration between these three member countries and continued high levels of inward FDI have further stimulated ongoing processes of economic liberalisation and trade openness in recent years.

During the period of import substitution, the process of industrialisation was geared primarily towards the domestic market. As a result of this, a large concentration of economic activity developed in and around Mexico City, which constituted the largest market in Mexico during this period (Krugman and Elizondo 1996). As a result, Mexico City developed rapidly into the 'urban giant' that is has become known for (Garza 1999). To indicate the speed of this process of agglomeration, during the period 1930-1960 the share of Mexico City in total manufacturing employment rose from 19 percent to 46 percent (Hanson 1997). There are several reasons for this rapid growth of Mexico's capital city. New Economic Geography (NEG) inspired explanations point at the importance for firms to locate in proximity to their main market (e.g. Krugman 1991, Krugman and Venables 1995, Fujita et al. 1999, Baldwin et al. 2003). By geographically concentrating their production, firms can reap the benefits from economies of scale. By locating in proximity to the main geographical concentration of consumers, firms are able to minimise on transport costs, as they can sell a large part of their production in the agglomeration. In a similar fashion, firms also generate important cost savings by locating in the region where other firms are located from which they can purchase inputs. As a result of this, a process of cumulative causation was set in motion that led to the rapid growth of an agglomeration in and around Mexico City. Other reasons for the rapid growth of this agglomeration of economic activity include the status of Mexico City as political and financial centre of the country, the implementation of favourable regional policies (Romero et al. 1952, Aguilar 1999) and the presence of strong push factors generating substantial migration flows from rural to urban areas (Alba 1982).

Trade Liberalisation and Changing Location Patterns

Following the introduction of trade liberalisation in the 1980s, the regional distribution of economic activity in Mexico started to change. Although all regions were of course affected by the implementation of the new development strategy, the effects on Mexico City and the states that share a border with the US have been most pronounced. Figure 4.1 shows the Mexican states. Table 4.1 presents indicators of the changes in the location pattern of manufacturing

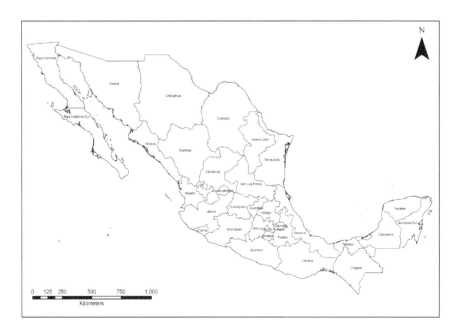

Figure 4.1   Mexican states

activity, measured as regional employment shares of Mexico City and the border states in total manufacturing employment during the period 1970-2003. Hanson (1996, 1997) reports on the break up of the traditional manufacturing belt for the period up until 1988, presenting evidence of a strong decrease in Mexico City's share in total manufacturing employment. The information presented in Table 4.1 indicates that this trend has clearly continued during the 1990s and early 2000s. Looking at the overall trend, in less than 25 years Mexico City has seen its share in manufacturing employment more than halved, from about 45 percent in 1980 to just over 20 percent in 2003. In contrast to this, the border states experienced a substantial increase in their participation in Mexico's manufacturing sector during the same period.

Table 4.1 also presents changes in the location pattern of more detailed manufacturing industries. The developments of the regional employment shares of these industries indicate that there is a substantial level of heterogeneity across the industries concerning the response to trade liberalisation. For instance, whereas Mexico City has largely maintained its share in industries such as food, beverages and tobacco, it has experienced large decreases in its share in industries of metal products and equipment, textiles and apparel and miscellaneous industries. The possibility of heterogeneity in industry response to decreasing international trade costs has also been identified in several theoretical studies. These studies look at the location of economic activity within a two-country framework, where one of the countries consists of a border region and an interior region. Krugman and Elizondo

Table 4.1   Employment shares in manufacturing industries: Mexico City and border states (%)

Mexico City							
Industries	1970	1980	1985	1988	1993	1998	2003
Manufacturing	41.9	44.4	36.8	33.2	29	23	21
Food, beverages, tobacco	22.8	28.7	26.6	25.6	23	23	21
Textile, apparel	51.4	43.9	36.9	33.6	27	10	19
Wood products	36.4	36.9	29	20.6	19	18	14
Paper, printing	52.6	65.1	57.6	54.3	48	45	41
Chemicals	60	55.7	43.6	45.6	46	40	40
Non-metallic minerals	33	29.8	29.2	29.1	20	11.6	16
Basic metals	21.2	34.6	28.3	26.1	27	15	19
Metal products and equipment	52.2	50.7	39.9	32	26	18	15
Other industries	60	69.2	55.1	55.1	41	30	21
Border States							
Manufacturing	n.a.	21	23.2	28.2	30	34	35
Food, beverages, tobacco	n.a.	17.7	18.9	18.5	19	19	19
Textile, apparel	n.a.	11.3	12.7	18	19	23	17
Wood products	n.a.	19.2	23.8	25.7	27	28	22
Paper, printing	n.a.	13.9	16	19	20	23	23
Chemicals	n.a.	14.6	14.4	16.6	21	23	24
Non-metallic minerals	n.a.	49.2	40.4	40.1	30	31	29
Basic metals	n.a.	32.1	28.6	31.8	32	44	42
Metal products and equipment	n.a.	26.9	34.3	46	50	58	59
Other industries	n.a.	15.6	22.6	30.6	29	40	53

Note: Mexico City is the Federal District and Estado de México; the border states are Baja California, Coahuila, Chihuahua, Nuevo León, Sonora and Tamaulipas.

Source: Hanson (1997) for the period 1970-1988; data from the Economic Census (1994, 1999, 2004) for other years.

(1996) find that economic activity agglomerates in the border region following the introduction of trade liberalisation. In contrast, other studies find that a decrease in international trade costs can also foster the agglomeration of economic activity in the interior region of the 2-region country (see Alonso-Villar 1999, Monfort and Nicolini 2000, Crozet and Koenig-Soubeyran 2004). Whether an industry agglomerates in the border region or the interior region depends on the relative importance of the main destination market.

This result offers an important explanation for the observed heterogeneity in industry response to the opening up of the Mexican economy. For example, the industry of metal products and equipment contains modern industries including the production and assembly of cars and car parts, televisions and personal computers. The strong change in location pattern of this industry, from being agglomerated in Mexico City to becoming agglomerated in the border states, can be explained by the fact that the US became the main destination market for most manufacturing firms in this industry.[1] In contrast, industries that have persisted to serve the domestic market, such as the industries of food, beverages and tobacco, have maintained a focus on Mexico City as main market, which explains why this industry has shown a continued strong geographical concentration in the capital city. In essence, as findings by Faber (2007) and also Jordaan and Sánchez-Reaza (2006) indicate, it is the export oriented industries that have become geographically concentrated in the border states; import competing industries have maintained a strong presence in the centre of the country. Overall, these important locational changes have altered the regional distribution of economic activity within Mexico from a situation with one main agglomeration in Mexico City to a situation where a majority of economic activity is concentrated in a limited number of regional production centres in the north and the centre of the country.

3. Evidence of Determinants of Regional Growth

The changes in the regional distribution of economic activity that followed the introduction of trade liberalisation have been substantial. Empirical research on the extent and nature of these changes can be classified broadly into two approaches. One approach is focused primarily on obtaining statistical evidence on the role of regional proximity to the market as determinant of the location of economic activity. The second approach analyses the nature of regional growth during the periods of import substitution and export promotion and tries to identify regional characteristics that have been important drivers of regional growth during these periods.

The Role of Proximity to the Market

Agglomeration of economic activity leads to the spatial variation of factor rewards. As Hanson (1997) explains, an agglomeration is characterised by higher wages, to compensate workers for congestion costs in the agglomeration (see also Brakman et al. 2001). Also, employees in an agglomeration can earn higher wages, caused by the increased productivity levels that they can generate

---

1   Another reason for the rapid growth of this industry in the border states is of course the growth of the maquiladora programme in the late 1980s and 1990s, in which many FDI firms of the sector of metal products and equipment are operating.

as a result of the presence of place-based externalities (Rosenthal and Strange 2004, Moretti 2004). The existence of such spatial variation of factor rewards can be used in empirical research on the identification of the importance for firms to locate in proximity to their main market. In essence, in the case of wages, the regional variation of factor rewards implies that there should be a negative relation between regional distance to the market and the regional wage level. Indications of the development of relative wages in the border states for the nine manufacturing subsectors are presented in Table 4.2, which is based on data presented by Jordaan and Sánchez-Reaza (2006).

Table 4.2   Relative wages in the border states

Industry	1980	1988	1993	1998	2003
Food, beverages, tobacco	0.776	0.882	0.899	0.827	0.904
Textile, apparel	0.857	0.917	0.771	0.995	0.981
Wood products	0.729	0.905	0.901	1.21	1.42
Paper, printing	0864	0.705	0.924	0.718	1.05
Chemicals	0.825	0.868	0.747	0.711	0.78
Non-metallic minerals	0.624	0.782	0.933	0.839	1.04
Basic metals	1.24	0.785	2.21	1.694	2.03
Metal products and equipment	0.743	0.812	0.686	0.813	1.11
Other industries	0.735	0.760	0.744	0.908	1.13

Note: Relative wages are nominal average wages in the Border States/nominal average wages in Mexico City.

Source: Jordaan and Sánchez-Reaza (2006).

At the start of the 1980s, wages in the border states were clearly lower compared to average wages in Mexico City, confirming the spatial variation in factor rewards. The exception is the subsector of basic metals, which has been an important and successful industrial activity in the border states of Nuevo León and Coahuila for a long time. Looking at the development of relative wages in the border states following the introduction of trade liberalisation, it is clear that relative wages in most subsectors have increased steadily during the 1990s and early 2000s, confirming the notion that the negative effect of distance to Mexico City has decreased in strength. Furthermore, it also appears that there is a growing positive effect from proximity to the US market. As Table 4.2 indicates, six of the nine subsectors report relative wages larger than one in 2003. Not only can this important switch in the relative level of regional wages be interpreted as evidence of the continued weakening of the effect of distance to the old agglomeration of economic activity, it may also be perceived as evidence that regional proximity to the US has started to outweigh the negative effect of regional distance to Mexico City.

Hanson (1997) presents further corroborating evidence for the period 1965-1988 on the importance of proximity to the market, by estimating regression models on determinants of relative regional wages in which regional distance to the US and to Mexico City are included as control variables. The empirical findings support the notion that proximity to the market is important, as the estimations produce estimated significant negative coefficients for both distance variables.[2] Evidence for a more recent period is presented by Chiquiar (2008), who uses information from the 1990 and 2000 population census to estimate wage determinants. One of the main findings is a clear identification of a significant positive effect on wage levels from regional exposure to international markets, confirming the importance of the US as destination market. Hanson (1998) estimates the effect of regional distance to the US on regional employment growth. For the period prior to trade liberalisation, the findings indicate that regional distance to the US is not significantly associated with regional employment growth. The regional presence of input-output linkages is positively associated with regional employment growth, which can be taken as evidence of the presence of positive agglomeration economies. For the period of trade liberalisation, regional distance to the US adopts a significant negative effect on regional employment growth, confirming the importance of this new destination market.

Faber (2007) conducts an empirical study on regional employment growth for the period 1993-2003, assessing the effect of export oriented versus import competing industries on regional growth. The findings indicate that those industries that are characterised by a larger increase in international trade than in national output have grown faster in the border states than in interior regions. Industries that are facing increased competition from imports have fared relatively well in interior regions, which offer a form of 'natural advantage', due to the relative large distance between these regions and the US border. In combination, these findings offer clear evidence of the heterogeneous nature of the industry response to trade liberalisation, determined primarily by whether firms are producing for the domestic or international market.

Finally, Jordaan and Sánchez-Reaza (2006) conduct regression analysis to identify determinants of regional employment shares in manufacturing industries for the period 1985-1998. Distance to Mexico City, distance to the US and the level of regional specialisation and diversity are the main explanatory variables in the estimated regression models. One important finding is that regional specialisation and diversity are both positively associated with regional employment shares during the entire period, indicating the presence of positive dynamic agglomeration economies.[3] Another important finding is that although regional distance to Mexico

---

2   See also Hanson (1996) for additional evidence on the effect of regional distance to Mexico City for the garment industry.

3   The interpretation of agglomeration economies as dynamic phenomenon argues that regional specialisation and diversity may foster place-based externality effects (Hanson 2001a, Quigley 1998). According to this interpretation, regional specialisation

City has a negative effect during the entire period, the magnitude of the effect decreases during the period of trade liberalisation. The estimated negative effect of regional distance to the US becomes significant and increasing in size during the 1990s, the period when the US became the new destination market for many manufacturing firms.

Nature and Determinants of Regional Growth

The other approach to study spatial dimensions of changes of the Mexican economy entails the analysis of regional growth regimes within the framework of absolute and conditional convergence. Studies on processes of regional growth during the period of import substitution all present evidence of absolute convergence (Juan-Ramón and Rivera-Batíz 1996, Chiquiar 2005, Rodríguez-Oreggia 2005, Rodríguez-Pose and Sánchez-Reaza 2002). The essence of regional growth during this period is that southern states, as well as some states in the centre of the country, experienced relative high growth in GDP per capita, compared to the border states and Mexico City. In strong contrast to this, the period following the introduction of trade liberalisation is characterised by high growth rates for the border states in particular, whereas most other states dropped below the average rate of regional growth (Rodríguez-Pose and Sánchez-Reaza 2002, Rodríguez-Oreggia 2005, Aroca et al. 2005, Chiquiar 2005). As a result, regional growth in the last two decades has been characterised by the existence of absolute divergence, fostering increasing inequality between states.

In extension of these findings, a number of studies provide evidence of factors that have acted as drivers of regional growth. For instance, Rodríguez-Pose and Sánchez Reaza (2005), Rodríguez-Oreggia (2007) and Esquivel and Messmacher (2002) conduct conditional convergence growth regressions and find that the regional endowment of human capital is an important factor causing differences in regional growth rates. In extension of this, Chiquiar (2005) presents more extensive evidence on the growth effects of a variety of regional characteristics, including physical and human capital, infrastructure and the share of large manufacturing firms in regional production. The explanation for the favourable growth rates of the border states under trade liberalisation is then that these states were relatively well endowed with these regional inputs, allowing them to take advantage of the changed economic environment. Despite the importance of these findings, it is striking that these regional growth studies have largely overlooked the potentially important roles of the development of several agglomerations of economic activity within Mexico and the large increase in inward FDI, especially when considering that both these phenomena have been central features of the period of trade liberalisation. The development of agglomerations of economic

and diversity foster the generation and transmission of knowledge spillovers, promoting sustained productivity advances of agglomerations through time (see e.g. Glaeser et al. 1992, Henderson et al. 1995, Henderson 1997, also Rosenthal and Strange 2004).

activity may have resulted in the generation of agglomeration economies, whereas the growing level of foreign participation may also have generated important direct and indirect growth effects. Given this gap in the literature, I will introduce and present the findings from a new empirical study on regional growth in Mexico in the next sections of this chapter, paying special attention to assessing how important regional growth effects from agglomeration and FDI have been.

4. Convergence/Divergence and Drivers of Regional Growth

To start the analysis of regional growth in Mexico during the period of trade liberalisation, I focus first on the process of absolute convergence/divergence. The concept of absolute convergence refers to the expectation that, within a neo-classical framework, regions in a country that start at a level below their steady state growth rate will experience per capita output growth at a rate higher than other regions. Capturing the growth process for a given period, the empirical translation of this proposition is as follows (see Barrow and Sala-i-Martin 1995, also Magrini 2004, Rey and Janikas 2005):

$$(1) \quad \frac{1}{t}\left(\log\left(\frac{Yi,t0+t}{Yi,t0}\right)\right) = \alpha - \left(\frac{1-e^{\beta T}}{t}\right) \log{(Yi,t0)} + \varepsilon it;$$

where t is the number of years, Yi,t0 is per capita GDP of region i at the start of the period and Yi,t0+t is per capita GDP of region i at the end of the period. The non-linear part of equation (1) can be transformed into a log linear expression, which gives:

$$(2) \quad \frac{1}{t}\left(\log\left(\frac{Yi,t0+t}{Yi,t0}\right)\right) = \alpha - \beta\,(\log Yi,t0) + \varepsilon it.$$

If the regional growth regime in a country is subject to absolute convergence, the estimated β-coefficient will carry a negative sign, indicating that regions with a higher level of GDP per capita have a smaller growth rate compared to regions with a lower level of GDP per capita. In the case of an estimated positive sign, the growth regime is characterised by regional divergence, which entails a growing difference in GDP per capita between leading and lagging regions.

Table 4.3 present the findings from estimating equation (2) for several time periods. For the entire period 1970-2004, there is no clear identifiable trend of either convergence or divergence between Mexican states. This is not surprising, given the drastic change in the development strategy that occurred during this period. Looking at the period of import substitution (1970-1985), the estimated β-coefficient carries a significant negative sign, indicating the process of absolute

Table 4.3   Estimates of absolute convergence/divergence

	Full period	Import substitution	Trade liberalisation	Early trade liberalisation	Post NAFTA liberalisation	Omitting outliers	Recent liberalisation	Omitting outliers
	1970-2004	1970-1985	1988-2004	1988-1993	1994-1998	1994-1998 (*)	1999-2004	1999-2004 (**)
α	0.021 (3.23)a	0.065 (8.58)a	-0.024 (1.58)c	-0.066 (2.10)b	-0.019 (0.90)	-0.03 (1.66)c	0.036 (4.16)a	0.046 (4.81)a
β	-0.002 (0.77)	-0.02 (5.59)a	0.013 (2.26)a	0.026 (2.10)b	0.01 (1.20)	0.016 (2.10)a	-0.003 (0.98)	-0.008 (2.00)b
R2	0.02	0.53	0.15	0.14	0.05	0.13	0.033	0.13

Notes: t statistics in parentheses; a, b and c indicate significance levels of 1, 5 and 10%.

Campeche and Tabasco are omitted from the sample, as their GDP is heavily influenced by the discovery of oil in the 1980s.

(*) Quintana Roo omitted.

(**) Distrito Federal and Nuevo León omitted.

convergence. In contrast, the estimated β-coefficient carries a significant positive sign in the period of trade liberalisation (1988-2004), confirming the presence of regional divergence. Having said so, additional estimates for sub-periods indicate that the process of regional divergence does not appear to apply to the entire period of trade liberalisation. For the period of early trade liberalisation (1988-1993), there is clear evidence of divergence. This is maintained during the period of post NAFTA liberalisation (1994-1998), once I omit the outlier state Quintana Roo from the sample. The findings for the most recent period of trade liberalisation (1999-2004) indicate however that the rate of divergence has slowed down. In fact, omitting the two outlier states the Federal District and Nuevo León gives findings that indicate the presence of a mild level of regional convergence.

In addition to these estimates of β-convergence, additional information on the regional growth regime can be obtained based on the concept of σ-convergence (Barrow and Sala-i-Martin 1995). This concept of σ-convergence focuses on the level of dispersion of income per capita between regions and can be calculated as the standard deviation of per capita output for the 32 Mexican states. Figure 4.2 presents the findings from calculating this indicator of dispersion for the period 1970-2004. The development of the level of dispersion of regional per capita output is in line with the findings obtained from the growth regressions. The level of dispersion clearly decreases between the years 1970-1985, when considering all states except the oil states of Campeche and Tabasco. Then, after the introduction

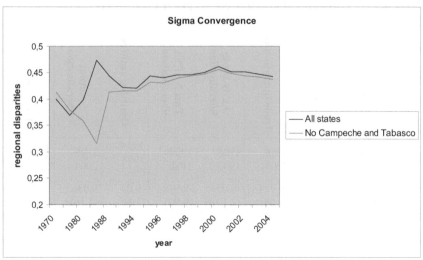

Figure 4.2   σ-convergence Mexican regions: 1970-2004

Source: Own estimations, based on data provided by INEGI.

of trade liberalisation, the level of dispersion increases steadily, reaching a high peak around the turn of the century. During the last years of the period, a small decrease in the level of dispersion sets in, which is in line with the findings of mild convergence for the last period of trade liberalisation as shown in Table 4.3.

Data and Variables

Although the findings on β- and σ-convergence provide important indications of the changes in regional growth in Mexico in the last two decades, they are based on the unrealistic assumption that all states share the same steady state per capita output level. In order to identify the nature of regional growth, whilst controlling for differences in steady state output levels, conditional convergence growth regressions are more appropriate. In terms of the regression model, this entails the following modification:

$$(3) \quad \frac{1}{t}\left(\log\left(\frac{Yi,t0 + t}{Yi,t0}\right)\right) = = \alpha - \beta\,(\log Yi,t0) + \sum_{i=1}^{j}\beta_j\,Xi,t0 + \varepsilon it;$$

where Xi,t0 is a vector of right hand side variables that capture differences in per capita steady state output levels. For the present analysis, I specify that this vector contains variables that capture the presence of agglomeration economies, the type and level of regional foreign participation and a selection of other regional characteristics that have been found in previous studies to cause differences in regional growth rates (see Chiquiar 2005, Rodríguez-Oreggia 2005, Rodríguez-Pose and Sánchez-Reaza 2005). To identify determinants of regional growth for the period of trade liberalisation, I built a regional dataset for the period 1988-2004. Within this period, I distinguish between three sub-periods of 1988-1993, 1994-1998 and 1999-2004. This gives a panel of 32 states with three time periods. However, similar to previous studies, I omit the oil states of Tabasco and Campeche from the sample, resulting in 90 observations.

The dependent variable is average regional growth of output per capita for the three time periods. The right hand side variables are listed in Table 4.4 with their definitions and sources. I include two variables that control for the regional variation of human capital. The first variable captures the regional level of schooling, measured as the average number of years of schooling of the regional economic active population. The second variable focuses on the regional variation in unschooled labour and is measured as the percentage of the regional population that is unable to read and write.

In addition to these two variables, I also add an interaction variable between a border states dummy and the schooling variable, to capture the skill premium in the border states associated with trade liberalisation. Feenstra and Hanson (1997) estimate the impact of the presence of FDI in Mexican regions during the period 1970-1988 and find that labour demand by foreign-owned firms caused a substantial increase in the level of wage differences between skilled and unskilled labour. The border states skill premium variable should control for this effect (see Chiquiar 2005).

The second set of control variables is related to infrastructure. In the selection of regional infrastructure variables, I am confined to the use of variables that are available for all three time periods, resulting in the selection of the three variables listed in Table 4.4. The variable railroad captures the size of the regional railroad network, electricity measures the regional presence of a good functioning electricity network and telephone density captures the regional variation in the presence and size of communication networks.

Next, I include two variables that control for the existing regional production structure, which may obstruct or delay the growth of new manufacturing industries. One variable captures the size of regional agriculture, measured as the share of agriculture in regional GDP. This variable controls for the possibility that a region with a large presence of agricultural activity will find it more difficult to develop new manufacturing activity. The second variable that I include captures the growth effect of the existing regional industry mix, which may also slow down or prevent the development of new manufacturing activity. The motivation to include this variable is that the use of highly aggregated regional data introduces the possibility that differences in regional growth rates are not related to structural differences between regions (e.g. differences in schooling or infrastructure), but instead simply reflect differences in the regional industry mix (Rigby and Essletzbichtler 1997, 2000, Jordaan 2008b). This issue may be important, as the existing industry mix may hinder the development of new manufacturing activity. To control for this, I

Table 4.4   Variables and data-sources

Variable	Definition	Data-source
Schooling	Average number of years of schooling economic active population	Provided by INEGI
Illiteracy	Percentage of population unable to read and write	1985 (Chiquiar 2005); 1993 population census 1990; 1998 extrapolated based on population census 1990 and 2000
Border* schooling	Border state dummy* schooling	
Railroads	Railroads/100 km2	1985 (Chiquiar 2005); 1993 and 1998 based on Anuario de Estadísticas Estatal, various years
Electricity	Percentage of households with electrical supply	1985 (Chiquiar 2005); 1993 population census 1990; 1998 extrapolated based on population census 1990 and 2000
Telephone	Telephones/100 persons	1985 (Chiquiar 2005); 1993 and 1998 Anuario de Estadísticas Estatal, various years
Agriculture	% share Agriculture in regional GDP	1985 (Chiquiar 2005); 1993 and 1998 provided by INEGI
Industry-mix	Predictor of importance of fast growing industries under import substitution (see text for elucidation)	Economic census; 1985, 1988, 1993, 1998
Federal investment	% share federal government investment in regional GDP	1985 (Chiquiar 2005); 1993 and 1998 provided by INEGI
Dummy 1999	Dummy variable to capture change in growth regime	
FDI flows	Total FDI flows/GDP	Economic census 1988, 1933, 1998 and data provided by INEGI
Maquiladora	Nr. of regional employees working for maquiladora firms	Sistemas de Cuentas Nacionales de México. INEGI, various years
FDI manufacturing share	% share FDI in total regional manufacturing employment	Unpublished data, provided by INEGI

Table 4.4 continued   Variables and data-sources

Variable	Definition	Data-source
Manufacturing density	$\sum u \dfrac{(employees\,man\,sec\,tor)\_county}{(squarekilometers)\_county}$ $u = \dfrac{(employees\,man\,sec\,tor)\_county}{(employees\,man\,sec\,tor)\_state}$	Economic census 1988, 1993, 1998
Services density	$\sum v \dfrac{(employees\,serv\,sec\,tor)\_county}{(squarekilometers)\_county}$ $v = \dfrac{(employees\,serv\,sec\,tor)\_county}{(employees\,serv\,sec\,tor)\_state}$	Economic census 1988, 1993, 1998
Population density	$\sum w \dfrac{(population)\_county}{(squarekilometers)\_county}$ $w = \dfrac{(population)\_county}{(population)\_state}$	Economic census 1988, 1993, 1998

calculate a predictor of the effect of the industry mix in the following manner. I calculate value added growth for 2-digit manufacturing industries at the national level for the periods 1985-1988, 1989-1993 and 1994-1998. For each region, I then multiply these national industry value added growth rates with 2-digit industry employment shares in total regional employment. After summing these, I obtain an employment-weighted growth predictor of the existing mix of regional manufacturing activity for the periods 1988-1993, 1994-1998 and 1999-2004.

Another control variable that I include captures the level of federal government investment, measured as the share of federal investment in regional GDP. Also, I add a dummy variable for the time period 1999-2004, to capture the change in the nature of the regional growth regime from divergence to mild convergence (or a decrease in dispersion), as identified in the previous section.

Next, I include variables that control for the presence of agglomeration economies. An often-used approach to capture the presence of place-based externalities is to use variables that capture the scale of regional activity. The size of regional industries can be used to capture the presence of localisation economies, whereas the size of total regional economic activity or population is often used to capture the presence of urbanisation economies (Moomaw 1981, Eberts and McMillen 1999, Rosenthal and Strange 2004). The problem with this approach is that these variables are very likely to contain endogenous elements, which will bias their estimated effect. Also, it is likely that such variables are related to other control variables, creating an additional problem of multicollinearity. To circumvent these problems, I follow an alternative strategy introduced by Ciccone and Hall (1996), which entails using the level of density of regional industries or total regional economic activity as indicators of agglomeration economies. I calculate density levels of the regional manufacturing sector, services and total regional population to capture localisation economies from manufacturing and services and urbanisation economies that are linked to overall regional economic activity. Important to consider here is that calculating these density scores at the state level is likely to produce ill-measured density indicators, as industries and population are not distributed uniformly within the states. To control for this, I calculate the three density indicators at the municipality level, measured as the number of manufacturing employees, services employees and population per square kilometre. I then aggregate these density scores for each state, using the municipalities' share in state level manufacturing employment, services employment and population as weights, which produces state-level density scores that are corrected for differences in density across municipalities in the states.

Finally, I add variables that capture the level and type of regional foreign participation. The first variable is total inward FDI flows as percentage of regional GDP, capturing the overall growth effect from total new inward FDI.[4] Second, I include the regional number of manufacturing employees working for maquiladora firms, to capture the growth effect of this specific type of FDI. Third, I calculate an indicator of the regional presence of total foreign-owned manufacturing activity, in the form of the ratio of the number of employees working for foreign-owned manufacturing firms over the total number of regional manufacturing employees. I use this indicator of the regional level of foreign participation to capture externality effects from manufacturing FDI (see Jordaan 2008b).

Empirical Findings

The main empirical findings from estimating a variety of specifications of regression model (3) are presented in Table 4.5. The first column present the findings from a random effects estimation of the regression model that contains all explanatory variables except the agglomeration and FDI variables. The estimated effect of initial GDP per capita is significant and negative, indicating the presence of conditional convergence of growth rates among Mexican states. The variable telephone density has an estimated positive effect, reflecting the positive nature of the growth effect of regional infrastructure. The variables that capture the regional importance of agriculture and the regional industry mix both carry significant negative coefficients, confirming the notion that these factors slow down or hinder the development of new manufacturing activity. The interaction variable of schooling with the border states dummy carries a significant positive coefficient, capturing the skill premium in the border states. The estimated

---

4   This concerns the total value of FDI inflows into all sectors of a region.

effect of the 1999 dummy is positive, confirming the structural break in growth regime. Also, the estimated effect of federal investment is positive, capturing the regional growth effect from this type of investment. Finally, the estimated effect of schooling, illiteracy and the infrastructure variables railroads and electricity are insignificant.

The next two columns contain the findings from adding the agglomeration and FDI variables to the regression model. Looking first at the growth effects from agglomeration economies, the findings indicate that there are two distinct effects. On the one hand, there appear to be positive localisation economies from both regional manufacturing and services, indicated by the positive estimated effect of the density scores for manufacturing and services. On the other hand, the estimated negative effect of density of population suggests the presence of negative urbanisation economies. Next, the estimated effect of regional foreign participation is also heterogeneous. The variable measuring the share of total inward FDI in regional GDP carries an insignificant coefficient. The estimated effect of regional maquiladora FDI is significant and positive, indicating that regions with a relative large presence of maquiladora FDI experience structurally higher growth rates. In contrast to this positive effect, the effect of the level of foreign participation in total regional manufacturing is negative, suggesting the presence of negative FDI externalities. One possible explanation for this estimated effect is the occurrence of a negative competition effect, where FDI firms steal market shares from Mexican firms (see Aitken and Harrison 1999), leading to decreased efficiency levels among these domestic firms. Another explanation may be that the presence of FDI firms drives up prices of regional inputs, which can lower the efficiency level of Mexican firms as well.[5]

Column four presents the findings from a feasible generalised least square (FGLS) estimation similar to Chiquiar (2005), using seemingly unrelated regression (SUR) that allows for correlated errors between panels (see Wooldridge 2002). There are not many differences with the findings presented in the previous columns. One difference is that in the SUR estimation none of the infrastructure variables carry a significant coefficient. Also, the estimated effect of illiteracy becomes significant negative, in line with expectations. The estimated effect of the agglomeration economies variables is stable, indicating the presence of positive localisation economies and negative urbanisation economies. As for the estimated effect of regional FDI, the variable capturing total inward FDI flows now carries a significant positive coefficient, indicating the positive overall growth effect from these FDI flows. The scale of maquiladora activity and the level of foreign participation in total regional manufacturing activity maintain their positive and negative effect.

---

5 These two effects are less likely to apply to maquiladora FDI. The negative competition effect from the presence of maquiladora firms is likely to be limited, as these firms produce mainly for the international market. Furthermore, if maquiladora firms use mainly low skilled, low wage labour, one can question to what extent this type of regional labour demand will generate substantial wage increases for Mexican firms.

Table 4.5　Determinants of regional growth: 1988-2004

	1	2	3	4	5
	Random effects	Random effects	Random effects	Feasible GLS	FGLS controlling for heteroscedasticity and autocorrelation
Initial GDP/ Cap	-0.0274 (0.0118)b	-0.05 (0.013)a	-0.0468 (0.0113)a	-0.0455 (0.011)a	-0.0468 (0.018)a
School	-0.035 (0.032)	-0.024 (0.039)	-0.027 (0.038)	-0.037 (0.028)	-0.027 (0.044)
Illiteracy	-0.0065 (0.009)	-0.0133 (0.01)	-0.0115 (0.01)	-0.02 (0.006)a	-0.011 (0.008)
Telephone	0.017 (0.0059)a	0.017 (0.004)a	0.0095 (0.05)c	0.0003 (0.0056)	0.009 (0.008)
Railroads	0.0001 (0.0008)	-0.0002 (0.0009)	-0.0006 (0.008)	-0.0005 (0.0006)	-0.0006 (0.00052)
Electricity	0.0031 (0.022)	-0.01 (0.024)	-0.0047 (0.021)	0.00067 (0.018)	-0.005 (0.02)
Agriculture	-0.008 (0.0039)b	-0.0117 (0.0042)a	-0.011 (0.0049)b	-0.0125 (0.0035)a	-0.011 (0.005)b
Federal investment	0.104 (0.025)a	0.096 (0.025)a	0.075 (0.029)a	0.068 (0.02)a	0.075 (0.026)a
Industry mix	-0.008 (0.0048)c	-0.011 (0.00148)a	-0.0143 (0.0039)a	-0.0152 (0.0033)a	-0.014 (0.004)a
School*border	0.0057 (0.003)b	0.009 (0.0038)a	0.01 (0.0032)a	0.008 (0.0025)a	0.01 (0.002)a
Dummy 1999	0.0255 (0.0044)a	0.0225 (0.004)a	0.028 (0.0053)a	0.029 (0.0035)a	0.028 (0.003)a
Density manufacturing		0.011 (0.0035)a	0.012 (0.0033)a	0.0128 (0.0024)a	0.012 (0.0018)a
Density services		0.0055 (0.0022)b	0.0061 (0.0027)b	0.0074 (0.0025)a	0.006 (0.0019)a
Density population		-0.0144 (0.004)a	-0.014 (0.0038)a	-0.015 (0.0027)a	-0.014 (0.0037)a
FDI/GDP			0.002 (0.0015)	0.003 (0.0009)a	0.002 (0.0007)a
Maquiladora employment			0.001 (0.0006)c	0.00098 (0.00045)b	0.001 (0.0005)a
Share FDI in regional manufacturing employment			-0.0049 (0.0014)a	-0.006 (0.0011)a	-0.005 (0.0017)a
LL	253.23	261.837	268.21	290.334	268.21
AIC	-470.44	-481.63	-488.42	-536.668	-492.42
Adjusted R2	0.52	0.60	0.66	0.66	0.67

Notes: Estimated standard errors in parentheses; a, b and c indicate significance level at 1, 5 and 10%.

Regressions 1 through 3 allow for clustered standard errors at the state level.

Regression 5: Test for heteroscedasticity (based on Drukker 2003): LR chi2(29) = 65.71 (prob.>chi2 = 0.001); Test for autocorrelation (based on Wooldridge 2002): $F(1,29)$ = 12.695 (prob.>F = 0.001).

Important to consider is that the method of SUR is based on the strong assumption that panels can only be related through their error terms. One could argue that this is a particularly strong assumption when using repeated observations through time. In light of this, I calculate test statistics to see whether the regression is affected by the presence of heteroscedasticity and autocorrelation, which appears to be the case (see notes to Table 4.5). Therefore, I re-estimate the model, controlling for both these issues. The findings are presented in the last column. Again, several of the right hand side variables that have been identified in previous research as having facilitated the change in growth regime between the periods of import substitution and trade liberalisation are not important when focusing on the second period only. In light of the insignificance of these variables, it is the more noteworthy that the agglomeration and FDI variables persist to be significantly associated with regional growth, constituting further support that both agglomeration economies and FDI effects have been important drivers of regional growth in Mexico during the last two decades. The feature that agglomeration economies and FDI have both positive and negative growth effects is also maintained.

Spatial Dimensions of the Effects of Agglomeration and FDI

The findings presented in Table 4.5 offer substantial support for the hypothesis that agglomeration economies and FDI effects have been important factors influencing regional growth in the Mexican economy. In this section, I extend the analysis to address the possibility that these effects can have spatial dimensions, as both the literature on agglomeration economies (Rosenthal and Strange 2004, Martin 1999, Parr 2002a, 2002b) and on FDI externalities (e.g. Smarzynska 2002, Driffield et al. 2004, Driffield 2006, Jordaan 2008b) argue that such spatial dimensions may be important and in clear need of further empirical testing and verification (see also Abreu, de Groot and Florax 2004, Henderson 2007). The findings presented in Chapter 4 on FDI location factors are in support of this, as they identify multi-regional dimensions of agglomeration economies.

To identify spatial dimensions of agglomeration economies and FDI effects, I need to augment the regression model with variables that capture potential spatial effects from these two sources. Doing so, the regression model becomes:

$$(4) \quad \frac{1}{t}\left[\log\left(\frac{Yi,t0+t}{Yi,t0}\right)\right] = = \alpha - \beta \, (\log Yi,t0) + \sum_{i=1}^{j}\beta_j \, Xi,t0 + \beta \sum_{k';1;r'\neq k}^{k} W \, Z + \varepsilon it;$$

where vector X is as defined earlier, vector Z contains variables that capture potential inter-regional effects from agglomeration and FDI and W is the distance matrix containing spatial weights w_{ik}, capturing the relation between geographical space and inter-regional effects from agglomeration and FDI.

I include all six agglomeration and FDI variables in vector Z. As I discussed in the previous chapter, I need to experiment with several specifications of the relationship between geographical distance and spatial effects, as it is not possible to determine *a priori* which specification is the most appropriate (Anselin 1988, Bode 2004). The need to try out several specifications becomes more apparent when considering that the various types of effect from agglomeration and FDI are likely to be generated by a variety of mechanisms. For instance, it is likely that localisation economies from manufacturing are transmitted via different mechanisms than general urbanisation economies are. Also, demonstration effects linked to the regional presence of FDI firms may be transmitted through different channels compared to FDI spillovers through input-output linkages. As each of these mechanisms may be affected by geographical distance differently, the analysis runs the risk of being too limited in scope when only one possible relation between geographical distance and spatial effects is considered.

I estimate regression model (4) with three alternative distance decay specifications. One is the first order contiguity assumption, which entails that spatial effects are assumed to possibly arise only between regions that share a border. In terms of the distance matrix W, the w_{ik} take the value of 1 when regions share a border and 0 otherwise. Second, I use the two nearest neighbours assumption, where spatial effects can arise between a state and its geographically two closest neighbouring states. The w_{ik} take the value of 1 for these states and 0 otherwise. Third, I use the inverse distance specification, where the spatial decay effect is assumed to increase continuously with increasing interregional distance (see e.g. Anselin 1988, Adserá 2000). In terms of the distance matrix W, the w_{ik} take the value of the inverse distance in kilometres between state capital cities.

The empirical findings from estimating regression model (4) are presented in Table 4.6. Column two presents the results from adding interregional effects to the regression model using the first order contiguity assumption. Importantly, although there are some minor changes in the magnitude of the estimated coefficients of the intra-regional agglomeration and FDI variables, the nature of their estimated effect is stable. As for inter-regional effects, both urbanisation and localisation economies from services appear to have a spatial dimension. Spatial urbanisation economies appear to be of a positive nature, whereas spatial localisation economies have a negative growth effect. Furthermore, all three variables capturing inter-regional FDI effects carry significant coefficients. The estimated positive inter-regional effect from total regional inward FDI indicates that the positive growth effect from overall FDI flows spills over across neighbouring regions. In contrast to this finding, the variables capturing spatial effects from maquiladora FDI and the level of foreign participation in regional manufacturing both carry a significant negative coefficient. An explanation for this finding may be that negative externalities from foreign participation transcend state borders. The negative competition effect may be multi-regional in nature, and it may also be the case that the upward pressure on regional input prices has an inter-regional dimension. Another explanation of these negative spatial growth effects can be found in the literature on growth effects from

Table 4.6   Spatial growth effects from agglomeration and FDI

	1	2	3	4	5
	No spatial effects	Contiguity 1	2 nearest neighbours	Inverse distance	No Mexico City
Initial GDP/Cap	-0.0468 (0.018)a	-0.05 (0.014)a	-0.055 (0.014)a	-0.058 (0.0018)a	-0.05 (0.01)a
School	-0.027 (0.044)	0.026 (0.03)	-0.003 (0.031)	0.045 (0.029)	0.06 (0.03)b
Illiteracy	-0.011 (0.008)	-0.024 (0.007)a	-0.0125 (0.006)a	-0.0165 (0.0055)a	-0.023 (0.008)a
Agriculture	-0.011 (0.005)b	-0.011 (0.004)a	-0.015 (0.003)a	-0.011 (0.0027)a	-0.018 (0.003)a
Federal investment	0.075 (0.026)a	-0.09 (0.06)	-0.0044 (0.054)	-0.004 (0.047)	0.035 (0.07)a
Industry mix	-0.014 (0.004)a	-0.011 (0.003)a	-0.01 (0.003)a	-0.012 (0.003)a	-0.02 (0.008)a
School*border	0.01 (0.002)a	0.004 (0.0039)	0.012 (0.005)b	0.011 (0.0045)b	0.002 (0.003)
Dummy 1999	0.028 (0.003)a	0.038 (0.0026)a	0.0365 (0.002)a	0.079 (0.0129)a	0.02 (0.01)b
Density manufacturing	0.012 (0.0018)a	0.011 (0.0025)a	0.011 (0.0032)a	0.011 (0.002)a	0.006 (0.0027)b
Density services	0.006 (0.0019)a	0.007 (0.0028)a	0.010 (0.003)a	0.0076 (0.0025)a	0.004 (0.003)
Density population	-0.014 (0.0037)a	-0.019 (0.0036)a	-0.018 (0.003)a	-0.0157 (0.0027)a	-0.021 (0.004)a
FDI/GDP	0.002 (0.0007)a	0.0017 (0.00037)a	0.0017 (0.0004)a	0.0017 (0.0006)a	0.001 (0.0009)
Maquiladora employment	0.001 (0.0005)a	0.0012 (0.00003)a	0.00176 (0.0004)a	0.001 (0.0003)a	0.0009 (0.00053)c
Share FDI in regional manufacturing	-0.005 (0.0017)a	-0.004 (0.001)a	-0.0045 (0.0009)a	-0.0048 (0.0014)a	-0.017 (0.008)b
Spatial variables					
Density manufacturing		-0.003 (0.006)	0.0128 (0.005)a	0.012 (0.019)	0.005 (0.008)
Density services		-0.029 (0.007)a	-0.0175 (0.005)a	-0.069 (0.009)a	-0.021 (0.009)b
Density population		0.0278 (0.009)a	0.0086 (0.10)	0.082 (0.028)a	0.028 (0.013)b
FDI/GDP		0.005 (0.0019)a	0.0024 (0.0012)b	-0.001 (0.001)	0.0065 (0.0015)a
Maquiladora employment		-0.0016 (0.001)c	-0.0002 (0.0012)	-0.0145 (0.007)b	-0.004 (0.001)a
Share FDI in regional manufacturing		-0.004 (0.001)a	-0.006 (0.0012)a	-0.0088 (0.004)b	-0.009 (0.0024)a

Table 4.6 continued   Spatial growth effects from agglomeration and FDI

	1	2	3	4	5
	No spatial effects	Contiguity 1	2 nearest neighbours	Inverse distance	No Mexico City
LL	268.21	273.24	278.37	273.05	279.84
AIC	-492.42	-490.487	-500.74	-490.09	-501.68
Adjusted R2	0.67	0.83	0.82	0.81	0.83

Notes: Estimated standard errors in parentheses; a, b and c indicate significance level at 1, 5 and 10%. Column 1 replicates column 5 of Table 4.5.

Columns 2 and 5 use first order contiguity assumption; column 3 is based on 2 nearest neighbours assumption; column 4 uses inverse distance between regions.

Distance matrices are row standardised.

investments in infrastructure. As shown by Boarnet (1998), public infrastructure investment in one region can have negative spillover effects on other regions, when the investment draws away investment flows from these other regions (see also Sloboda and Yao 2008). In a similar fashion, the estimated negative inter-regional growth effects from maquiladora FDI and foreign participation in the regional manufacturing sector may reflect a situation where FDI in one region draws away investment flows from neighbouring regions.

Columns three and four present the empirical findings from estimating regression model (4) with the other two specifications of the distance decay effect. Again, the estimated growth effects of the intra-regional agglomeration and FDI variables remain stable. The estimated spatial effects of the agglomeration economies variables are sensitive to the specification of the distance matrix. Using the two nearest neighbours specification produces a significant positive effect of localisation economies from manufacturing. The estimated effects of localisation economies from services and of urbanisation economies are insignificant. Using the inverse distance between regions specification produces findings similar to the results obtained with the first order contiguity assumption. As for the spatial effects from regional FDI, the estimated effects of total inward FDI flows and the regional scale of maquiladora activity are also sensitive to the distance decay specification. The two nearest neighbours assumption produces a significant positive spatial effect of total regional FDI inflows, whereas the negative effect from the regional scale of maquiladora activity is insignificant. Using the inverse distance between regions as spatial weighting variable produces an insignificant spatial effect of the FDI inflows variable and a significant negative effect for the maquiladora variable. The estimated negative spatial growth effect from the level of foreign participation in the regional manufacturing sector is significant in both estimations.

Finally, I re-estimate regression model (4) on a restricted sample that omits Mexico City. One could argue that, compared to the other regions, Mexico City constitutes a unique region, which may have affected the estimations with the full sample of regions. For instance, Mexico City is the number one destination region for new FDI. Also, the level and type of agglomeration economies in this urban giant may be structurally different from agglomeration economies in the other Mexican states. The findings from estimating the regression model on the restricted sample are shown in column 5. Compared to the findings in the last column indicate that the regression model appears to function particularly satisfactory, as most non-agglomeration and non-FDI variables, including schooling and illiteracy, carry significant coefficients with expected signs. As for the effect of intra-regional agglomeration economies, the findings indicate the presence of positive localisation economies from manufacturing and negative urbanisation economies. Also, the positive growth effect from the regional scale of maquiladora activity and the negative effect from the level of foreign participation in regional manufacturing are confirmed. Furthermore, the estimated spatial effects from agglomeration and FDI are similar to the findings obtained for the full sample. Overall, except for some differences in significance of non-agglomeration and non-FDI variables, there is a large degree of similarity between the findings from the full and the restricted sample, indicating the robustness of the findings on intra- and inter-regional growth effects from agglomeration and FDI.

5. Summary and Conclusions

In the 1980s, the Mexican government substituted a new development strategy of economic liberalisation and trade promotion for the unsuccessful strategy of import substitution and government intervention, generating drastic changes in the economic environment. The introduction of trade liberalisation also fostered important changes in the location pattern of economic activity, developing from a situation where Mexico City constituted the main agglomeration of economic activity to a situation where manufacturing activity has become concentrated in a limited number of regional production centres in the north and centre of the country. The main explanation for these distinct spatial changes is that the opening up of the Mexican economy made the border states attractive for many manufacturing firms, as the US became the new destination market for these firms. A second important development has been the dramatic increase in inward FDI into the Mexican economy, in response to policies that facilitate and actively attract new foreign-owned manufacturing firms.

A number of studies provide evidence on the nature and causes of the spatial changes in the Mexican economy. Studies on determinants of regional wages or regional employment shares confirm the importance of proximity to the market; also, they identify a decreasing importance of Mexico City and an increasing importance of the US as main market for many firms and entire industries. More detailed evidence shows that the border states have become important locations for export oriented industries, whereas import competing industries have maintained a strong presence in Mexico City. Regional growth studies present evidence that the introduction of trade liberalisation led to a structural change in the nature of regional growth. Whereas during the years of import substitution regional growth was characterised by the existence of absolute convergence, the change in development strategy led to the occurrence of absolute divergence. As for determinants of regional growth, the evidence is biased strongly towards identifying the effect of 'traditional' regional characteristics, including physical and human capital and infrastructure. The potentially important roles of agglomeration economies and FDI effects have received only little attention in these studies, which is the more remarkable given that the development of new agglomerations of economic activity and the large influx of new FDI have been central features of the period of trade liberalisation.

The empirical analysis of this chapter addresses this gap in the literature. One initial finding of the analysis is that it appears that regional divergence reached its peak in the late 1990s. Findings for the early 2000s indicate a decrease in the level of dispersion of GDP per capita and contain evidence of the presence of a mild level of absolute convergence. In addition to this, the conditional growth regressions indicate that the Mexican regions are subject to a process of conditional convergence, once I control for the growth effects of a variety of regional characteristics.

Next, the findings indicate that there are a number of regional characteristics that have acted as drivers of regional growth during the last two decades. Having said so, the estimations also indicate that several regional characteristics that have been identified in previous research as being important do not appear to be important for regional growth under trade liberalisation. An explanation for this difference in findings is that although these regional characteristics, most notably indicators of several types of regional infrastructure, have been related to the structural break in growth regime between the periods of import substitution and trade liberalisation, they have not played such an important role in regional growth processes within the latter period. In contrast to this, most of the agglomeration and FDI variables are significantly associated with regional growth, indicating that agglomeration economies and FDI effects have indeed played important roles as drivers of regional growth during the period of trade liberalisation.

Importantly, the findings also indicate that agglomeration and FDI generate both positive and negative growth effects. Looking at the effects from agglomeration, the findings suggest the presence of positive localisation economies from both manufacturing and services and negative urbanisation economies. As for the intra-regional growth effects from FDI, the regional level of total inward FDI flows and the regional scale of maquiladora activity are both positively associated with regional growth. The estimated negative growth effect from the level of foreign participation in regional manufacturing suggests that there are negative externality effects, caused by a negative competition effect or an upward pressure on regional input prices, which can both result in a decrease in efficiency among Mexican firms.

In extension of these findings, the analysis also addresses the existence and nature of inter-regional growth effects from agglomeration and FDI. Using several specifications of the relation between spatial effects and geographical distance, the findings suggest the presence of positive spatial urbanisation economies and negative localisation economies. As for the spatial growth effects from FDI, the estimations identify a positive effect from the overall regional level of inward FDI flows. Interestingly, the estimated spatial growth effects from the regional scale of maquiladora activity and from the level of foreign participation in regional manufacturing are both negative. This may indicate that the presence of FDI is generating negative spatial externalities, where the negative competition effect or the upward pressure on prices of regional inputs transcends state borders. Alternatively, the estimated negative spatial growth effects from these two FDI variables may reflect a situation where the presence of FDI in a given region draws away investment from other regions, which would show up as a negative spatial FDI growth effects in the estimations. Irrespective of the exact explanation, these findings further corroborate the notion that both agglomeration and FDI have played important roles as drivers of regional growth in Mexico in the last two decades, generating positive and negative growth effects at both the intra- and inter-regional level.

Chapter 5

Intra-industry FDI Externalities in Mexican Manufacturing Industries: Endogeneity, Technology Gap and Agglomeration

1. Introduction

The last two decades have witnessed an impressive growth of the body of empirical research on intra-industry FDI productivity effects in host economies. An initial surge of cross-sectional studies produced evidence of the occurrence of FDI externalities of a positive nature. Findings from more recent studies, often estimating FDI effects in panel data settings, are much more diverse, ranging from positive to insignificant to negative externality effects. As a result, opinions on this type of FDI externalities differ markedly. In response to this impasse, several studies have started to include assessments of possible determinants of these FDI effects. In particular, several studies have focused on indentifying the effect of the level of absorptive capacity of domestic firms in a host economy, by relating the level of technological differences between FDI and domestic firms to the occurrence of FDI spillovers. More recently, another line of investigation has started to develop on the relation between geographical space and FDI externalities, in an attempt to capture a broader range of FDI spillovers and to investigate whether these FDI effects are influenced by geographical dimensions.

Against this background, the purpose of this chapter is to develop an empirical analysis on intra-industry FDI spillovers in the Mexican economy. Several previous studies have estimated FDI effects in Mexican manufacturing industries and provide important evidence on the existence and nature of FDI spillovers. However, this evidence is characterised by two important limitations. One limitation is that the evidence is based on data for the 1970s and 1980s. For most of these years, the Mexican economy was subject to strict policies of import substitution and government intervention. The empirical analysis in this chapter is based on data for the 1990s, making the analysis more relevant for understanding FDI effects in the changed economic environment of economic liberalisation and trade promotion. Second, evidence on the effect of the technology gap between FDI and Mexican firms on FDI spillovers is conflicting and the effect of geographical proximity on FDI effects has received only little attention. In contrast to this, one of the main foci of the analysis in this chapter is to obtain statistical evidence on the effect of these two determinants on the occurrence and perhaps also the nature of FDI spillovers in Mexican manufacturing industries.

The chapter is constructed as follows. In section 2 I survey the main empirical evidence on FDI externalities in the Mexican economy. In section 3 I discuss the main features of the cross-sectional dataset that I built, based on unpublished data from the 1994 Mexican manufacturing census. Section 4 addresses the core criticism of estimations of FDI externalities with cross-sectional data that estimated FDI effects may be affected by problems of endogeneity. In particular, the criticism holds that ordinary least squares (OLS) estimates of FDI externalities are biased upwards, caused by a tendency among FDI firms to gravitate towards high productivity industries. In this section, I discuss this important problem in detail and introduce a new instrumental variable that allows me to obtain unbiased estimates of FDI externalities. Section 5 presents the main findings on the effects of the technology gap between FDI and Mexican firms and of geographical proximity between these two types of firm on FDI externalities. I use both a dataset with national industry level observations and a dataset for several important individual Mexican states to identify the effects of these two determinants. Finally, section 6 summarises and concludes.

2. Intra-industry FDI Externalities in Mexico: A Survey of the Evidence

Several empirical studies have addressed the occurrence of FDI externalities in the Mexican manufacturing sector. Ramirez (2000) conducts a time series study, estimating the relation between the stock of foreign capital in the Mexican economy and aggregate labour productivity for the period 1960-1995. The findings show a significant positive relation between changes in the (lagged) foreign-owned capital stock and estimated labour productivity, which Ramirez interprets as evidence of the presence of positive FDI productivity effects. In a related study, Ramirez (2006) extends on these findings by estimating the effect of FDI capital stock that is corrected for profit and dividend remittances flowing out of the Mexican economy. Although the magnitude of the estimated positive association between labour productivity and foreign owned capital stock decreases somewhat, the significance of the positive relationship is maintained, in support of the impression that FDI generates positive externalities.

The majority of the evidence on FDI externality effects has been obtained from the type of empirical study surveyed in Chapter 2. One set of empirical studies is based on the analysis of unpublished cross-sectional industry-level data from the 1970 Mexican manufacturing census. This set of papers was initiated by Blomström and Persson (1983) and Blomström (1989), who regress an indicator of labour productivity of the Mexican-owned share of a set of manufacturing industries on several control variables, including the capital-labour ratio of Mexican firms in these industries, an indicator of human capital, a measure of the level of market concentration and an indicator of the industry level of foreign participation, measured as the share of FDI firms in industry employment. The estimated effect of this FDI participation variable is significant and positive, which is interpreted as evidence of the occurrence of positive FDI externalities. In a related study,

Blomström (1986) presents further corroborating evidence of positive FDI effects. In this study, Blomström estimates the effect of industry foreign participation in a regression model where the dependent variable captures the extent to which average labour productivity approaches best practice industry productivity, finding that a high level of FDI participation leads to a smaller intra-industry deviation from best practice productivity.

Kokko (1994, 1996) uses the same dataset to obtain evidence on the effect of the technology gap on positive FDI spillovers. Estimating regression models largely in line with Blomström and Persson (1983) and Blomström (1989), Kokko (1994) identifies a group of manufacturing industries that do not benefit from positive FDI externalities. The findings indicate that the estimated effect of industry foreign participation remains insignificant in those industries that are characterised by large technological differences between FDI and Mexican firms and a relative large level of industry foreign participation. Kokko labels these industries as 'enclaves', where large technological differences between the two types of firm prevent the materialisation of positive FDI effects (Kokko, 1994, p. 291). In Kokko (1996), the empirical analysis focuses on identifying externality effects from the level of competition between FDI and Mexican firms, as well as from industry foreign participation. The findings indicate that positive externality effects arise from both these sources, but only when the analysis omits the 'enclave' industries.

Finally, Blomström and Wolff (1994) estimate the impact of industry foreign participation on labour productivity growth of Mexican manufacturing firms and on the rate of convergence in labour productivity levels between FDI and Mexican firms for the period 1970-1975. Again, the findings are in support of the notion that FDI firms generate positive externality effects, as both labour productivity growth and the rate of convergence are positively associated with the industry level of foreign participation. Interestingly, the findings of this study contrast findings presented by Kokko (1994). Whereas Kokko (1994) finds that large technological differences between Mexican and FDI firms prevent positive externalities to materialise, Blomström and Wolff (1994) find that the initial level of technological differences between the two types of firm actually has a positive effect on the level of productivity growth of Mexican firms and the rate of productivity convergence between FDI and Mexican firms. This suggests that large technological differences are conducive rather than detrimental to positive FDI externalities.

The second set of studies on FDI spillovers is based on the analysis of plant level data for the 1980s. The evidence from these studies on FDI effects is more mixed. Aitken et al. (1996) investigate whether industry foreign participation generates wage spillovers. Controlling for the effects of capital, technology acquisitions and industry and regional effects, Aitken et al. (1996) estimate whether the industry presence of FDI is significantly associated with industry wage levels. Estimations for aggregate wage levels indicate a significant positive effect of industry foreign participation. However, when estimating the effect on the industry wage level of only Mexican firms, the estimated effect becomes insignificant, suggesting the absence of FDI wage spillovers.

Using the same dataset, Aitken et al. (1997) estimate for the presence of another type of FDI spillover effect in the form of market access spillovers. To test for the presence of this type of spillover, Aitken et al. (1997) estimate determinants of the likelihood that a Mexican plant is engaged in exporting activities. The empirical findings indicate the presence of positive FDI externalities, as the likelihood that a Mexican plant is engaged in exporting is positively associated with the level of regional concentration of FDI exporting activity. There is no significant effect of the level of regional concentration of overall exporting activity, supporting the importance of the regional presence of FDI exporting firms. The importance of geographical proximity for this type of FDI spillover effect is underlined by the finding that the level of foreign participation in industry exports at the national level does not affect the probability that a Mexican firm is selling on international markets.

Finally, Grether (1999) uses the dataset for the 1980s to identify FDI externalities by estimating determinants of the deviation of plant level efficiency from the efficiency level of best practice plants. The findings show a significant negative association between the industry level of foreign participation and the relative efficiency level of Mexican plants, indicating that industries with a high participation by FDI firms are characterised by a relative large gap between productivity of Mexican plants and best practice plants. Grether (1999) interprets this as evidence that FDI hinders the diffusion of technology, which does not necessarily mean that negative FDI externalities are at play though. A negative association between industry FDI participation and relative efficiency of Mexican plants may indicate the presence of negative externalities, if the presence of FDI firms lowers productivity levels of Mexican plants via the occurrence of a negative competition effect (see Aitken and Harrison 1999). However, the estimated negative association may alternatively indicate that FDI firms are more successful in lowering or preventing the occurrence of positive externalities in those industries that are characterised by a relative high level of foreign participation.

3. Data, Regression Model and Definition of Variables

Although it is of course difficult to compare the findings from these previous studies, due to differences in methodology, data and regression models, overall there is substantial evidence that the presence and operations of FDI firms in Mexican manufacturing industries are linked to externalities of an intra-industry nature. However, these findings apply to a period when the Mexican economy was characterised by policies of import substitution and government intervention. Especially in light of the drastic changes in development strategy towards economic liberalisation and trade promotion that have occurred from the late 1980s onwards, the existing evidence on FDI effects from these studies may have become less relevant to understand effects from FDI in such a changed economic environment. Furthermore, the findings on the effect of the technology gap on FDI spillovers are conflicting and only one study presents findings that suggest that geographical proximity between firms may be important for these FDI effects.

The dataset that I built for the empirical analysis of this chapter is based on unpublished data provided the Instituto Nacional de Estadística y Geografía (INEGI), the main government institute in Mexico responsible for carrying out and disseminating findings from the main economic censuses and surveys. INEGI is prevented by law to disseminate plant level information from the manufacturing census, to prevent the possibility that individual plants can be identified. As a compromise, INEGI agreed to aggregate plant level observations from the 1994 manufacturing census, containing observations for 1993, to detailed 6-digit national level manufacturing industries. This aggregation was carried out separately for Mexican and foreign-owned plants. As such, the nature of the dataset is very similar to the dataset used by the set of papers initiated by Blomström and Persson (1983) and Blomström (1989). The dataset contains, for FDI and Mexican plants separately, the following industry level variables: value added, total fixed assets at book value, number of white collar employees, number of blue collar employees and number of manufacturing employees. The distinction between Mexican and foreign-owned observations means that for instance for the variable value added, the dataset contains for each industry an observation of the magnitude of value added for the aggregation of Mexican plants and for the aggregation of foreign-owned plants in the industry separately.[1] After inspection of the data cells and deleting industries with missing information on value added or number of manufacturing employees, the dataset contains observations for 228 6-digit manufacturing industries that can be used for the empirical analysis.

Production Function and Definition of Variables

Externalities can not be measured directly and therefore have to be identified in an indirect manner. The common approach in applied economics research on FDI externalities is to estimate some form of production function for domestic firms or industry shares in a host economy, whereby the level of intra-industry foreign participation is included as one of the control variables that may affect productivity of these firms or industry shares. An estimated significant association between productivity and the industry level of foreign participation is interpreted as evidence of the occurrence of FDI externalities, with the sign of the association indicating the nature of the externalities. Given the nature of the dataset, I derive the regression model from a standard Cobb-Douglas (CD) production function. The standard CD function can be depicted as follows:

(1) $Q = A * K^{\alpha} * L^{\beta};$

---

[1] In line with OECD definitions, all plants in the underlying database that have 10 percent or more of total assets under foreign control are considered to be foreign owned.

where Q, K and L are production, capital and labour, α and β are the output elasticities of capital and labour and A represents the efficiency parameter. Estimating regression (1) with cross-sectional data is usually problematic, given difficulties with accurately measuring capital and the likely presence of multicollinearity between the main variables. Given these issues, it is better to express the CD function in its intensive form (see Intrilligator et al. 1996), depicting production per unit of labour as a function of the capital labour ratio as follows:

(2) $Q = A*(K/L)^{\alpha} * L = A * (k)^{\alpha} * L$, where $(k) = K/L$;

dividing both sides by L gives the production function in its intensive form:

(3) $Q/L = A*(k)^{\alpha}$.

Stating equation (3) in log linear form gives the equation that can be estimated directly with the available data:

(4) $\ln(Q/L)_{(m)} = \ln A + \alpha \ln(k)_{(m)} + \varepsilon$;

where m stands for Mexican-owned industry share and ε is the error term.

Equation (4) states that labour productivity of the Mexican-owned share of an industry can be depicted as a function of the capital labour ratio of the Mexican-owned industry share and the efficiency parameter A, which contains the effects of other factors that can be hypothesised to influence the estimated level of labour productivity. For the present analysis, I specify that the efficiency parameter contains the effects of human capital, market concentration, agglomeration economies and intra-industry foreign participation. The variables are listed in Table 5.1 with definitions and data-sources. Table 5.2 present summary statistics.

I measure the dependent variable as the ratio of value added over the number of manufacturing employees in the Mexican-owned share of an industry. The capital labour ratio is measured as the ratio of total assets at book value over the number of manufacturing employees for the Mexican owned industry share. To measure the level of human capital, I follow Blomström and Persson (1983) by taking the ratio of the number of white collar employees over the number of blue collar employees in the Mexican-owned industry share.

It is likely that the industry productivity level is affected by the industry level of market concentration (e.g. Kokko 1994). It is not clear what the exact nature of this effect is, though. A positive effect would indicate that the level of concentration enhances productivity, indicating that firms are successfully engaged in some level of monopoly pricing, thereby raising value added. Alternatively, a negative effect would suggest that productivity is fostered by competition. To measure the level of market concentration I use a Herfindahl index, which captures the combined effects of the influence of the number of plants, market shares and coalition potential (see

Table 5.1 Variables and definitions

Name	Description	Definition
(Q/L)(m)	Labour productivity Mexican owned share of industry	$$\frac{(valueadded)(m)}{(numberofemployees)(m)}$$
k(m)	Capital labour ratio Mexican owned share of industry	$$\frac{(totalassetsatbookvalue)(m)}{(numberofemployees)(m)}$$
LQ(m)	Human capital Mexican owned share of industry	$$\frac{(numberofwhitecollaremployees)(m)}{(numberofbluecollaremployees)(m)}$$
Herfi	Market concentration	$$\sum_{i=1}^{n}\left(\frac{xi}{X}\right)^2$$ where x = 1…n is the set of plants in the industry; xi is production of plant i, X is total industry production (*)
Gini	Geographical concentration	$$\left(\frac{\sum_{i-1}^{n}\sum_{j=1}^{n}\left|xi-xj\right|}{2n^2\mu}\right)$$ where n = number of states (32); xi, xj = 1…n are the state shares in industry employment; μ is average state industry share
For	Intra-industry foreign participation	$$\frac{(numberofemployees)(f)}{(numberfomployees)(m)+(numberofemployees)(f)}$$

Notes: m = Mexican owned industry share; f = foreign-owned industry share; (*) = The census only provides average plant production per plant size class of an industry. To proxy the Herfindahl index depicted in Table 5.1, I take the average plant share in total industry production for all the size classes. I then take the squared value of the average plant share of a size class and multiply it by the number of firms in the size class. Doing so for all size classes and aggregating gives the proxy for the level of market concentration.

Source: (Q/L)(m), (k)(m), LQ(m) and For are calculated with unpublished data provided by INEGI; Herfi and Gini are calculated with published data from the 1994 Economic Census. All data are 1993 observations.

Table 5.2   Summary statistics

Variable	Mean	Std dev	Min	Max
$(Q/L)_{(m)}$	3.719	0.618	2.155	5.681
$k_{(m)}$	3.776	0.984	1.083	5.957
$LQ_{(m)}$	-1.136	0.538	-2.548	1.386
Herfi	2.95	2.346	-4.024	7.971
Gini	-0.236	0.167	-1.07	-0.0317
For	-1.952	1.364	-5.981	-0.011

Note: Nr of observations = 228; all variables are in logs.

van Lommel et al. 1977). This index is calculated as the aggregation of squared shares of individual plants in industry production.

I include an indicator of the nature of the geographical distribution of the industries to control for the presence of agglomeration economies. The need to do so is supported by findings for Mexico by Grether (1999) that indicate that the relative efficiency of Mexican plants is positively affected by agglomeration economies. Also, the findings presented in Chapter 4 indicate the presence of agglomeration economies at the regional level. To capture agglomeration economies, I calculate a Gini-coefficient of the distribution of the manufacturing industries over the 32 Mexican states. Although usually applied to indicate the level of inequality of income distributions (see Owell 1977), the Gini coefficient can also be used to obtain an indication of the type of geographical distribution of industries within a country (see e.g. Krugman 1991). The variable Gini ranges between the extreme values of 0 and 1, where 0 indicates that all 32 Mexican states have an equal share in an industry and 1 indicates that an entire industry is geographically concentrated in one state.

Finally, I include a variable that captures the intra-industry level of foreign participation. Following the approach taken by most empirical studies on FDI productivity effects, I measure the industry level of foreign participation as the ratio of the number of manufacturing employees working for FDI firms over the total number of manufacturing employees in an industry.

4. OLS versus IV Estimations of FDI Externalities in Mexican Manufacturing Industries

An important advantage of the dataset is that the 6-digit industry observations are based on an all-inclusive census among all manufacturing plants in Mexico. Furthermore, the dataset contains information for a more recent period compared to previous research, which will make the findings more relevant for understanding the operations and effects of FDI in an economic environment of liberalisation and trade promotion. Having said so, the cross-sectional nature of the dataset also brings with it some limitations. Cross-sectional estimations produce associations between variables, with is not the same as identifying causal relationships. Also, the data does not allow for an explicit analysis of the time dimension of FDI externalities, and there is the issue that estimated associations may be affected by biases from aggregation and omitted variables. Finally, and most importantly, there is the problem of causality or endogeneity, which can bias the estimated externality effect of FDI (see Hanson 2001b, Aitken and Harrison 1999).

To get a clear understanding of the problem of endogeneity, the main regression model that I estimate to identify FDI productivity effects is:

$$(5.A) \quad (Q/L)_{(m)} = \beta 0 + \beta 1 \, k_{(m)} + \beta 2 \, LQ_{(m)} + \beta 3 \, \text{Herfi} + \beta 4 \, \text{Gini} + \beta 5 \, \text{For} + \varepsilon.$$

The focus of the estimation lies on the coefficient of FOR. For instance, if the estimation, controlling for the effects of the other control variables, produces a significant positive association between Mexican productivity and the industry level of foreign participation, this can be interpreted as evidence of the occurrence of positive externalities. However, this conclusion rests on the assumption that the line of causation of the estimated association runs from FOR to productivity of Mexican firms. The problem is that it is often very difficult to predict the line of causation *a priori* (Keller 2004). Although an estimated association between the two variables is informative on the nature and strength of the relationship between the variables, it does not tell us that the assumption of the regression model that the line of causation runs from FOR to labour productivity is correct. The strong criticism of the early cross-sectional evidence of FDI spillovers centres exactly on this issue of causality. In particular, the criticism holds that evidence of positive externalities in the form of a positive association between industry foreign participation and productivity of domestic firms or industry shares is likely to have been caused by a tendency among FDI firms to gravitate towards high productivity industries in a host economy (Aitken and Harrison 1999, Hanson 2001b). Applying this critique to equation (5.A), an estimated positive association between FOR and labour productivity of Mexican-owned industry shares would then be caused by a tendency among FDI firms to concentrate in high productivity industries, rather than reflecting the case that Mexican firms are subject to positive FDI externalities. In such a case, the variable FOR is endogenous to equation (5.A), resulting in a biased estimation of the effect of FOR on productivity. If the estimated positive association between FOR and Mexican productivity contains both the productivity effect of foreign participation on Mexican firms and the effect that FDI firms gravitate towards high productivity industries, it becomes impossible to distil the unique FDI externality component from the estimated association.

Instrumenting FOR

Empirical studies that benefit from the use of panel data are able to apply various solutions to this problem of causality. Solutions include the use of lagged values for the FOR variable, regressing the change of productivity of domestic firms on industry foreign participation or the use of instruments based on lagged values of other right hand side variables. These solutions are not available to cross-sectional studies. In a cross-sectional setting, the problem of endogenous FDI requires the use of instrumental variable estimation techniques (Keller 2004).

A suitable instrument needs to meet two key criteria (see Wooldridge 2002). First, the instrument must be exogenous to the regression model. Second, the instrument must be sufficiently correlated with the endogenous right hand side variable. In other words, '…a good instrument is correlated with the endogenous regressor for reasons that the researcher can verify and explain, but uncorrelated with the outcome variable for reasons beyond its effect on the endogenous regressor' (Angrist and Krueger 1999: 8). Finding such an instrument is not an easy task. For the present analysis, it means that I need to find a variable that is sufficiently associated with the variable FOR, whilst at the same time this variable should not affect productivity of Mexican firms, other than through its association with FOR. There are of course several variables that are related to the industry level of foreign participation. For instance, if foreign-owned firms are attracted to fast growing industries, there will be a positive association between the industry level of foreign participation and industry fixed capital formation (see Kholdy 1995). Also, it is very likely that there will be a substantial association between the industry level of profitability and industry foreign participation (see Dunning 1993). The problem with these variables is that they do not meet the criterion of not being associated with productivity of Mexican firms, other than through their association with the variable FOR, as both fixed capital investment and profitability are linked to labour productivity in more ways than only through their association with FOR.

For the analysis in this chapter, I propose to use an instrumental variable in the form of the general variation of FDI intensity across manufacturing industries. Data presented in OECD (2002) provides a first indication of this variable. Figure 5.1 shows the percentage share of FDI in the total number of manufacturing employees for several manufacturing industries of a group of OECD member countries. It is clear that there is substantial variation in the level of foreign participation across the industries. For instance, chemical industries, petrochemical products and electrical machinery are characterised by a level of foreign participation of more than 35 percent. In contrast, fabricated metal products and wood and paper products show much lower levels of foreign participation of around 10 percent. The importance of this industry variation of foreign participation is that it indicates that there appears to be a general difference in FDI intensity between these manufacturing industries. This means that some of the variation of industry foreign participation across industries in any host economy will be caused by this general difference

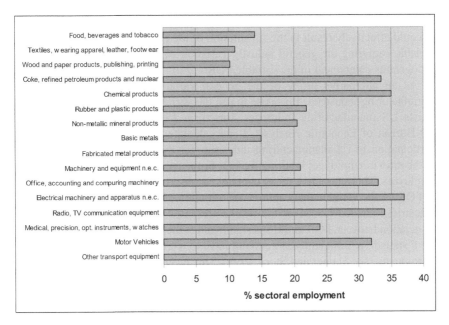

Figure 5.1   Industry foreign participation: OECD countries, 1990s

Note: Simple averages. The data covers 19 OECD countries; available years differ across countries.

Source: Own calculations, based on data presented in Table VI.5. 'Percentage share of employment in foreign affiliates in selected industries OECD average, 1990s'. In chapter 6 'Trends in Foreign Direct Investment in OECD countries', *OECD Economic Outlook*, no. 73, 2002. Paris: OECD.

in FDI intensity. In other words, irrespective of the host economy under analysis, it is likely that the industry of fabricated metal products will have a lower level of foreign participation than the industry of electrical machinery, caused by this general difference in FDI intensity that exists across manufacturing industries.

However, the information presented in Figure 5.1 suffers from two weaknesses when it comes to calculating a suitable instrumental variable for the variable FOR in regression model (5.A). One issue is that there are only 16 observations, which is likely to lower the strength of the association with the variable FOR, raising the problem of weak instruments (see Bound, Jaeger and Baker 1995). Second, there is a problem caused by the fact that the majority of FDI in Mexico is US owned, in combination with the feature that the US is also the major source of outward FDI in OECD countries. This means that US FDI is responsible to a considerable extent for the variation of FDI participation over the manufacturing industries as shown in Figure 5.1. If the distribution of US FDI over the OECD industries is affected by the productivity levels of these industries, the use of the variation of industry foreign participation over these industries as instrument for the variable

FOR in regression model (5.A) does not solve the problem of endogeneity. Instead, it would effectively result in the substitution of the endogenous component of the variable capturing the general industry variation of foreign participation for the endogenous component of the variable FOR.

As an alternative proxy of the general variation of FDI intensity across manufacturing industries, I propose to use the variation of average FDI intensity across US manufacturing industries. Based on freely available data from the US Bureau of Economic Analysis, I can calculate the average level of foreign participation in US manufacturing industries for the period 1988-1995, defined as the ratio of employees working for FDI firms over the total number of industry employees. This alternative proxy contains 52 observations. A bivariate regression of FOR on the variable that captures the variation of FDI participation across US industries, labelled US, gives the results presented in Table 5.3.[2] The estimated association between FOR and US is significant at the 1 percent level. Furthermore, the F statistic is sufficiently large to not have to worry about problems with weak instruments (see Staiger and Stock 1999). This means that the variable US meets the first criterion of a good instrument of being sufficiently associated with the variable FOR.

Table 5.3   Regression of FOR on US

Constant	US	Adj. R2	F value	N
-0.95 (0.23)a	0.42 (0.09)a	0.10	19.43 (0.00)	228

Note: Estimated standard errors in parentheses; robust to heteroscedasticity using Hubert/White/Sandwhich method. a indicates significance level of 1%.

The second requirement is that the instrument can not be associated with labour productivity of Mexican firms, other than through its association with the variable FOR. One reason why it is very likely that the instrumental variable US meets this requirement is that the vast majority of FDI in the Mexican economy is US owned. A problem could arise if FDI into US manufacturing industries and into Mexican industries from third countries is related. A company that wants to service the US market may decide to locate production in Mexico to benefit from labour cost advantages and service the US market from Mexico. Also, it could be that FDI inflows into American and Mexican industries are complementary. In these cases, MNEs may consider characteristics of both American and Mexican industries, creating the possibility that the cross-industry variation of FDI participation over the industries in both countries are related to industry characteristics that in turn are related to productivity levels of Mexican industries. These problems do not

---

2   See Jordaan (2004a) for a description of how US and Mexican manufacturing industries can be matched.

arise in the present case, given the strong dominance of FDI from the US in the Mexican economy.[3] Another issue that would prevent the use of the variable US as instrument in the present analysis would arise if Mexico has a substantial share in US inward FDI. Suppose I want to estimate regression model (5.A) for the UK, the country with the largest share in US inward FDI stock in the 1990s.[4] It may be the case that UK outward FDI towards the US originates from those industries in the UK to which US MNEs are attracted. In such a case, the variation of UK-owned FDI over US manufacturing industries will share an endogenous component with the distribution of US FDI over UK manufacturing industries. This problem does not arise in the present analysis, as Mexico has a negligible share in US inward FDI. In 1998 for instance, less than 0.3 percent of FDI stock in the US was under the control of Mexican-owned MNEs.[5]

It appears that the variable US can serve as a good instrument for the variable FOR in regression model (5.A). Prior to applying the instrument to estimate FDI externalities, the first required step is to determine whether the variable FOR is indeed endogenous to the regression model. This can be assessed by means of a Hausman specification test (Hausman 1978). To perform the test, I assume the following equation system:

(5.A.a) $(Q/L)_{(m)} = \beta 0 + \beta_x X + \beta_{for} FOR + \varepsilon_a;$

(5.A.b) $FOR = \gamma 0 + \gamma (Q/L)_{(m)} + \gamma_{us} US + \varepsilon_b$

$X = (k_{(m)}, LQ_{(m)}, Herfi, Gini).$

The idea is to regress the variable FOR on the exogenous variables of equation (5.A.a) and the variable US, as shown in (5.A.b). From this, I obtain the estimated residuals of (5.A.b), which I add to equation (5.A.a). If ε_b carries a significant coefficient in the estimation of (5.A.a), the variable FOR must be treated as endogenous to the regression model, warranting the use of instrumental variable estimation. A summary of the findings of the second stage regression are shown in Table 5.4. The results show that the residuals ε_b from equation (5.A.b) are significantly associated with labour productivity of Mexican firms in equation (5.A.a). This confirms that the variable FOR is endogenous to the original regression model, which means that OLS estimation will produce a biased estimate of the effect of FOR on labour productivity of Mexican firms. To identify the extent and nature of this bias, I compare OLS and instrumental variable (IV) estimations of regression model (5.A) in the next section.

---

3   In more recent years, there has been a substantial increase in the level of third country FDI into the Mexican economy. However, in 1993 the vast majority of FDI in Mexico was US owned.

4   Based on data taken from www.bea.gov.

5   Ibid.

Table 5.4   Summary of findings second stage regression

εb	Adj. R2	F	N
-0.19 (0.08)a	0.70	81.68 (0.00)	228

Note: Estimated standard errors in parentheses; robust to heteroscedasticity using Hubert/White/Sandwhich method. Adjusted R2 and F statistic are for estimating equation (5.A.a) to which residuals εb have been added. a indicates significance level of 1%.

FDI Externalities in Mexican Manufacturing Industries: Comparing OLS and IV

Table 5.5 presents the main findings from estimating regression model (5.A) with OLS and IV estimation techniques.[6] The first column with results contains the findings from estimating the regression model with standard OLS techniques. The estimation is characterised by satisfactory statistics. The adjusted R^2 is 68 percent, indicating that a substantial part of the variation of the dependent variable is explained by the control variables. Furthermore, looking at the right hand side variables besides FOR, the estimated effects are as expected. The estimated positive effect of the capital labour ratio and human capital indicate that Mexican firms with relative high levels of capital intensity and human capital have structurally higher labour productivity levels. The estimated positive effect of the Herfindahl index of market concentration indicates that more concentrated industries have higher labour productivity. The estimated effect of the Gini variable is also positive, reflecting the presence of positive agglomeration economies.

Looking at the main variable of interest, the OLS findings suggest the presence of positive FDI externalities. The estimated coefficient of the variable FOR is positive and significant, which can be interpreted as evidence that Mexican firms are subject to positive FDI spillover effects. The magnitude of the coefficient is modest, however, indicating that a 10 percent increase in the industry level of foreign participation generates a 0.4 percent increase in labour productivity. However, we know from the findings of the Hausman specification test that the variable FOR is endogenous to the estimated regression model. This means that the estimated effect of this variable is biased. Moreover, following the criticism of this type of cross-sectional estimation, the estimated effect of FOR is biased upwards, as foreign-owned firms can be assumed to gravitate towards high productivity industries (Aitken and Harrison 1999, Hanson 2001b).

The second column of results presents the findings from the instrumental variable estimation, where the variable FOR is instrumented with the variable US. The F statistic of the first stage regression is sufficiently large and the Wu-Hausman test statistic confirms the endogeneity of the variable FOR.

---

6   For related findings, see Jordaan (2004a, 2004b).

Table 5.5   FDI externalities in Mexican manufacturing industries: Comparing OLS and IV findings

Variables	OLS	IV_1	IV_2
Constant	2.85 (0.18)a	3.11 (025)a	3.10 (0.25)a
$K_{(m)}$	0.30 (0.03)a	0.33 (0.07)a	0.33 (0.07)a
$LQ_{(m)}$	0.26 (0.06)a	0.15 (0.08)c	0.19 (0.07)a
Herfi	0.08 (0.01)a	0.05 (0.025)b	0.05 (0.024)b
Gini	0.63 (0.13)a	-0.95 (0.62)	-0.71 (0.56)
FOR	0.04 (0.012)a	0.27 (0.13)b	0.22 (0.11)b
Adj. R^2 first stage	0.68	0.41	0.44
F first stage	92.07 (0.00)	25.95 (0.00)	22.42 (0.00)
Wu-Hausman F test	–	4.61 (0.03)	4.05 (0.05)
Adj. R^2 second stage	–	0.55	0.59
F second stage	–	55.05 (0.00)	55.55 (0.00)
Hansen J statistic	–	–	1.246 (0.26)
N	228	228	228

Note: Estimated standard errors in parentheses; heteroscedasticity robust using Hubert/White/Sandwhich method. a and b indicate significance levels of 1 and 5%. IV_1 uses US as instrument; IV_2 uses US and US_VA as instruments.

However, comparing the findings from the OLS and the IV estimations, there is no evidence that the OLS estimation produces an upward bias in the estimated effect of industry foreign participation. In fact, the IV estimation produces a significantly larger estimated effect of this variable. This indicates that, instead of producing an upward bias, the use of OLS produces a downward bias in the estimated intra-industry externality effect of FDI. Looking at the findings from the IV estimation, the industry presence of FDI appears to generate a much larger positive effect on labour productivity of Mexican firms: an increase in the industry level of foreign participation by 10 percent leads to a 2.7 percent increase in labour productivity. The difference in findings between the OLS and IV estimations indicates that, contrary to the common assumption that FDI firms are attracted to high productivity industries, in the present case FDI firms are attracted towards low productivity industries. Controlling for this bias, the

IV estimation produces unbiased findings that constitute robust evidence of the occurrence of substantial positive FDI externality effects.

One potential problem with the findings obtained with the IV estimation is that the instrumental variable US itself may contain an endogenous element. If foreign investment in general is attracted to industries with particular productivity levels (high or low), the use of the instrument US may substitute this general endogeneity problem for the endogenous component of the variable FOR. To control for this potential problem, I calculate a second instrument, labelled US_VA, in the form of the variation of average value added over US manufacturing industries for the period 1988-1995. Adding this extra instrument controls for any possible relation between the variable US and the variation of productivity levels across US industries. The findings from adding this second instrument are presented in the last column of Table 5.5. The Hausman test statistic confirms the endogeneity of the variable FOR and the F statistic of the first stage regression is sufficiently large. The Hansen J statistic indicates that both instruments are accepted. The estimated coefficient of industry foreign participation has decreased somewhat in comparison with the IV_1 estimation, but remains positive and significantly larger than originally obtained via OLS estimation.

The finding that the unbiased estimation of FDI externalities with IV techniques produces a larger positive coefficient than with the OLS estimation challenges the common criticism of cross-sectional estimates of FDI effects, as the findings indicate that FDI firms gravitate towards low rather than high productivity industries. I see three explanations for this particular type of causation running from low labour productivity to high FDI industry presence. First, efficiency seeking FDI that locates in Mexico is very likely to gravitate towards industries with relative low levels of labour productivity, as these industries are characterised by labour intensive production technologies. FDI that locates in Mexico to benefit from low labour costs will concentrate in these low productivity, labour intensive industries, causing the negative line of causation from industry productivity to industry foreign participation.

Second, the existence of the maquiladora programme undoubtedly has a strengthening effect on this particular relation between industry labour productivity and industry foreign participation. It is well-know that labour intensive maquiladora industries have shown persistent low levels of value added (Ramirez 2003). MNEs are stimulated to participate in these low value added industries of the Mexican manufacturing sector, benefitting from special tax breaks on imported inputs and re-exported assembled products and relying on the use of labour intensive production technologies. As a result, the negative line of causation from the industry productivity level to industry foreign participation is further strengthened.

Third, also market seeking FDI firms may actively choose to locate in low productivity industries. It may be the case that FDI firms follow a strategy to maximise market shares in the Mexican economy. Such a strategy may result in foreign-owned firms gravitating towards low productivity industries, if the low productivity level of Mexican firms is a reflection of their level of competitiveness. Furthermore, following the idea that the success of FDI hinges upon the successful protection and maintenance of ownership-specific advantages (see Dunning 1993, also Kugler 2006), foreign firms may decide to locate in industries with large technological differences between FDI and Mexican firms, resulting in an overrepresentation of FDI firms in manufacturing industries with relative low productivity levels among Mexican firms.

Summing up, this section has produced two important findings. First, I have introduced a new instrument that can be used to obtain unbiased estimates of FDI externalities in a cross-sectional setting, as the proxy for the general variation of FDI intensity across manufacturing industries meets both criteria of a good instrument. Furthermore, I establish that the variable capturing the intra-industry level of foreign participation is indeed endogenous to the regression model, warranting the use of IV estimation. Second, the instrumental variable estimations produce findings that are in strong contrast to the common criticism of cross-sectional estimations of FDI externality effects. This criticism entails that OLS estimations of FDI spillovers are likely to be biased upwards, caused by a tendency among FDI firms to concentrate in high productivity industries in a host economy. In contrast to this, the findings for Mexico show that the OLS estimation actually *underestimates* FDI externalities. This can be explained by a tendency among FDI firms to gravitate towards manufacturing industries with relative low productivity levels, a feature that is very likely to apply to FDI in the Mexican manufacturing sector. As a result, the IV estimations produce evidence of a significantly larger positive effect of intra-industry FDI participation on labour productivity of Mexican firms, indicating that Mexican firms are enjoying positive FDI externalities.

5. Determinants of FDI-Externalities: Technology Gap and Agglomeration

The large heterogeneity in empirical findings on FDI effects has stimulated research on structural factors that may influence the occurrence and perhaps also the nature of FDI spillover effects. As I discussed in Chapter 2, there are several of such factors that may be important, including the nationality of FDI firms (Haskel et al. 2007, Javorcik et al. 2004), the underlying motivation of FDI (Driffield and Love 2007), whether FDI firms are fully or partially foreign owned (Javorcik and Spatareanu 2003, Sjöholm and Blomström 1999) and several host economy characteristics (Blomström and Kokko 2003, Crespo and Fontoura 2007). In this section, I focus the attention on two potentially important factors in the form of the level of technological differences between FDI and domestic firms and the industry level of agglomeration. The importance of obtaining more evidence on the effect of the technology gap between FDI and Mexican firms is that, as discussed in Chapter 2, the evidence on the effect of the technology gap is not clear cut and there are problems with the assumed link between the technology gap and the concept of absorptive capacity. As for the effect of the industry level of agglomeration, the high level of similarity between channels that transmit FDI spillovers and mechanisms that are linked to the creation and transmission of agglomeration economies strongly suggests that industry agglomeration will have a positive effect on FDI externalities. As discussed in Chapter 2, this important hypothesis has received only very limited attention in the literature thus far.

One approach to identify effects of the technology gap and the industry level of agglomeration on FDI spillovers is to split the sample of industries into two groups, whereby industries are distinguished based on their level of technological differences between FDI and Mexican firms or their level of agglomeration. For instance, to identify the effect of the technology gap between FDI and Mexican firms on FDI spillovers, I divide the industries into two groups, based on the industry level of technological differences between the two types of firm. I then estimate regression model (5.A) for the two groups of industries and compare the estimated effect of industry foreign participation between the two groups of industries.

One indicator of the level of technological differences that I use is the industry-wide level of technological complexity. Following Kokko (1994), I measure the industry level of technological complexity as the ratio of total assets at book value of FDI firms over the total number of manufacturing employees of these firms, assuming that the level of capital intensity of FDI firms is positively related to the level of technological complexity of an industry and that a high level of technological complexity implies large technological differences between FDI and Mexican firms. The second indicator of the technology gap between the two types of firm that I use is in line with Blomström and Wolff (1994), measured as value added per manufacturing employee in FDI firms divided by value added per manufacturing employee in Mexican firms. This indicator is based on the assumption that a large difference in labour productivity between FDI and Mexican firms is reflective of large technological differences between the two types of firm. The third indicator that I use to classify the industries into two groups is based on the industry level of geographical concentration. To distinguish between industries that are agglomerated to either a high or a low extent I use the variable Gini, defined as in the previous section.

The empirical findings from estimating regression model (5.A) separately for the groups of industries are presented in Table 5.6. The first two sets of columns with findings show the results from estimating regression model (5.A). for industries that are characterised by a high or low level of technological differences between FDI and Mexican firms, using either technological complexity or the technology gap as distinguishing factor. Following the traditional interpretation that technological differences are direct inversely related to the level of absorptive capacity of domestic firms, positive FDI externalities should be particularly visible in industries with a small technology gap. However, the empirical findings show exactly the opposite: positive FDI spillovers occur in industries with a large technology gap. This finding is in

Table 5.6   Effects of the technology gap and agglomeration on FDI externalities

	Technological complexity		Technology gap		Agglomeration	
Variable	*Low*	*High*	*Small*	*Large*	*Low*	*High*
Constant	2.94 (0.27)a	3.18 (0.26)a	3.53 (0.25)a	2.85 (0.21)a	2.88 (0.28)a	3.23 (0.27)a
$k_{(m)}$	0.29 (0.05)a	0.35 (0.05)a	0.42 (0.09)a	0.45 (0.06)a	0.29 (0.05)a	0.29 (0.05)a
$LQ_{(m)}$	0.15 (0.10)	0.33 (0.07)a	0.23 (0.06)a	0.21 (0.06)a	0.22 (0.06)a	0.22 (0.10)b
Herfi	0.10 (0.02)a	0.05 (0.02)b	0.28 (0.06)a	0.36 (0.06)a	0.07 (0.01)a	0.08 (0.022)a
Gini	0.86 (0.13)a	0.56 (0.19)a	0.26 (0.05)a	0.11 (0.05)b	0.48 (0.16)a	0.94 (0.63)
For	0.09 (0.08)	0.23 (0.06)a	0.08 (0.06)	0.11 (0.05)b	0.08 (0.06)	0.20 (0.10)b
Adj. R^2	0.76	0.57	0.74	0.71	0.82	0.58
F	63.17 (0.00)	34.52 (0.00)	60.55 (0.00)	58.54 (0.00)	81.32 (0.00)	37.34 (0.00)
N	98	128	107	119	93	153

Notes: Estimated standard errors in parentheses; robust to heteroscedasticity using Hubert/White/Sandwhich method. a and b indicate significance levels of 1 and 5%.

For each of the three potential determinants, I have split the sample into two groups of industries, based on several cut off points such as the mean and median value of the determinants. Chow test statistics indicate that the differences between the groups of industries as shown in the table are significant. All findings are obtained via instrumental variable estimation, using the specification of the IV_2 regression in Table 5.5.

direct contrast to the absorptive capacity hypothesis, and also contradicts earlier findings for Mexico by Kokko (1994). Instead of reflecting the level of absorptive capacity, the alternative interpretation of the technology gap, as I propose in Chapter 2, is more appropriate to explain the present findings. Large technological differences reflect the presence of a large potential scope of positive externalities. This large scope of potential externalities offers incentives to Mexican firms to engage in externality facilitating investments, raising their level of absorptive capacity. Also, industries with large technological differences are less likely to suffer from negative competition-induced effects, as FDI and Mexican firms are less likely to be in direct competition with each other. The last set of columns contains the findings from estimating the regression model for industries that are geographically concentrated to either a high or a low extent. The findings support the notion that agglomeration enhances positive FDI spillovers. Comparing the findings between the two groups of industries, the estimated effect of industry foreign participation is significant and positive in those industries that are agglomerated to a high extent. This suggests that geographical proximity between Mexican and FDI firms fosters positive FDI externalities, caused by the positive effect of agglomeration on the existence and the workings of channels of these FDI spillovers.

To obtain further evidence on the effects of the technology gap and agglomeration on intra-industry FDI spillovers and to identify FDI externality effects that are cleared from the influence of these two determinants, I estimate an augmented version of the regression model, adding interaction variables between the variable measuring intra-industry foreign participation and the technology gap and industry agglomeration. This gives the following regression model:

(5.B) $(Q/L)_{(m)} = \beta_0 + \beta_1 k_{(m)} + \beta_2 LQ_{(m)} + \beta_3$ Herfi $+ \beta_4$ Gini $+ \beta_5$ FOR $+ \beta_6$

FOR*Tech $+ \beta_7$ FOR*Gap $+ \beta_8$ FOR*Gini $+ \beta_9$ FOR*Tech*Gap $+$

β_{10} FOR*Tech*Gini $+ B_{11}$ FOR*Gap*Gini $+ \varepsilon$;

where Tech stands for the industry level of technological complexity, Gap is the technology gap between FDI and Mexican firms and Gini is the level of industry agglomeration.

The results from estimating regression model (5.B) are shown in Table 5.7. The first column with results contains the findings from estimating the regression model with all the interaction variables. The interaction variable between FOR and Gap carries a significant positive coefficient, indicating that industries with a large presence of FDI firms and large technological differences between FDI and Mexican firms are subject to positive FDI spillovers. Again, this finding is in contrast to Kokko (1994), who labelled industries with a large technology gap and large FDI presence as 'enclave industries', industries that did not benefit from positive FDI effects in the 1970s. As the present findings indicate, the situation appears to have been completely reversed following the introduction of trade liberalisation, as it is these enclave industries that are subject to positive FDI externalities. None of the other interaction variables carry a significant coefficient however, and the estimated effect of intra-industry foreign participation that is cleared from the effects of technological differences and industry agglomeration is also no longer significant.

Of course, using all the interaction terms simultaneously in the regression model is likely to introduce the problem of multicollinearity. Therefore, I have re-estimated less augmented versions of the regression model, containing the interaction variable between Gap and FOR and one of the other interaction variables in turn. The findings presented in the second column with results indicate that there is an additional effect from the determinants of FDI externalities. In addition to the positive effect of the technology gap on FDI spillovers, the interaction term between FOR, Gap, and Gini also carries a significant coefficient. This suggests that geographical proximity does enhance

Table 5.7   Interaction effects between intra-industry foreign participation and determinants of externalities

	Full model	Only significant interaction effects
Constant	3.01 (0.26)a	3.23 (0.17)a
$k_{(m)}$	0.29 (0.04)a	0.28 (0.03)a
$LQ_{(m)}$	0.27 (0.07)a	0.25 (0.05)a
Herfi	0.09 (0.01)a	0.09 (0.015)a
Gini	0.11 (0.07)	0.85 (0.17)a
FOR	0.03 (0.11)	0.13 (0.06)b
GAP*FOR	0.04 (0.01)a	0.03 (0.007)a
GINI*FOR	-0.27 (0.27)	–
TECH*FOR	0.003 (0.02)	–
GINI*GAP*FOR	0.06 (0.04)	0.05 (0.027)b
GINI*TECH*FOR	-0.04 (0.03)	–
Adj. R^2	0.72	0.74
F	63.24 (0.00)	80.21 (0.00)
N	228	228

Note: Estimated standard errors in parentheses; robust to heteroscedasticity using Hubert/White/Sandwhich method. FOR is instrumented as in column IV_2 of Table 5.5. a and b indicate significance levels of 1 and 5%.

positive FDI externality effects, but only in those manufacturing industries with a sufficiently large technology gap. In addition to this finding, the estimated effect of the variable FOR is now significant again with a positive sign, indicating that the effect of intra-industry foreign participation, cleared from the effects of the technology gap and industry agglomeration, is the creation of positive FDI externalities.

Intra-industry FDI Externalities at the Regional Level

The estimations using national level manufacturing industries produce findings that indicate that FDI firms are generating positive externalities. Furthermore, the results indicate that determinants in the form of the technology gap and industry agglomeration both stimulate these externality effects. In addition to the analysis of national level industries, the analysis of FDI effects at the regional level can produce further indications of the importance and nature of FDI externalities and the effects of the determinants. This is especially so when considering that FDI externalities may be more pronounced at the regional level, given the positive relation between geographical proximity between firms and spillovers.

To estimate FDI spillovers at the regional level, I obtained additional unpublished data from INEGI from the 1994 manufacturing census for several individual Mexican states. In selecting states for the analysis, I have focused on states that have played an important role in the Mexican economy during both the period of import substitution and trade liberalisation and states that have rapidly gained importance following the introduction of trade liberalisation. States that have played an important role during both import substitution and trade promotion are Estado de México, Jalisco and Nuevo León.[7] The states that have experienced substantial growth under trade liberalisation are the maquiladora states Baja California, Chihuahua and Tamaulipas. The variables that the datasets for these states contain are similar to the dataset for national industries described earlier. I estimate the augmented regression model for these states, with two modifications. One modification is that I measure the level of industry agglomeration different. For the analysis among individual states, I measure the industry level of agglomeration at the state level in the form of a location coefficient, defined as the ratio of the share of a regional industry in total regional manufacturing employment over the share of the national industry in total national manufacturing employment. Second, instead of capturing only the effect of intra-industry foreign participation at the national level, I now also control for the intra-industry foreign participation effect at the regional level. This gives the following empirical model:

(5.C) $(Q/L)_{(m)} = \beta_0 + \beta_1\, k_{(m)} + \beta_2\, LQ_{(m)} + \beta_3\, \text{Herfi} + \beta_4\, \text{Gini} + \beta_5\, \text{FOR-national} +$

$\beta_6\, \text{FOR-regional} + \beta_7 \text{FOR*Tech} + \beta_8\, \text{FOR * Gap} + \beta_9\, \text{FOR*Gini} +$

$\beta_{10}\, \text{FOR*Tech*Gap} + \beta_{11}\, \text{FOR*Tech*Gini} + B_{12}\, \text{FOR*Gap*Gini} + \varepsilon;$

---

7   Mexico City consists of the Federal District and Estado de México. The data for the Federal District may be unreliable, as many firms assign employment and production to their headquarters in the District, whereas actual production sites may be located elsewhere. For this reason, I take Estado de México to represent Mexico City.

FOR-national captures the industry variation of FDI participation across national-level industries, measured as the ratio of manufacturing employees working for FDI firms over the total number of manufacturing employees in a national manufacturing industry.[8] FOR-regional is defined similarly, but calculated for manufacturing industries of the individual states. The interaction variables between intra-industry foreign participation and the technology gap and industry agglomeration are all calculated with the FOR-regional variables.

The main findings from estimating regression model (5.C) for the six states are presented in Table 5.8. The first column with findings is from estimating the regression model on the pooled sample of states. Comparing the findings with those for the national industries, the estimated effects of the control variables are similar in nature, with one important difference. Intra-industry foreign participation at the national level generates positive FDI externalities, as indicated by the positive coefficient of the FOR-national variable. In contrast to this, the estimated effect of intra-industry foreign participation at the regional level is significant negative. As such, these findings are in line with Chapter 4, where the level of regional foreign participation is negatively associated with regional growth. This indicates that the regional dimension of intra-industry FDI externalities results in the materialisation of negative externalities. An explanation for this finding may be that the negative competition effect from the presence of FDI firms is particularly pronounced at the intra-regional level. Alternatively, the estimated regional dimension of negative FDI externalities may be caused by an increase in prices of regional inputs, caused by the presence of FDI firms. The findings also contain several indications that the determinants of externalities are important. The interaction variables between Gap and FOR and Tech and FOR carry significant positive coefficients, which indicates that industries with large technological differences enjoy additional positive FDI externalities. Furthermore, the interaction variable between Gini, Gap and FOR also carries a significant positive coefficient, confirming the positive effect of geographical proximity on positive FDI externalities, provided that the level of technological differences between FDI and Mexican firms is sufficiently large.

The findings for the states Estado de México, Jalisco and Nuevo León represent externality effects in those Mexican states that have incorporated substantial shares of the Mexican manufacturing sector during both the periods of import substitution and trade promotion. Looking at the estimated effect of intra-industry foreign participation, Estado de México and Jalisco benefit from positive FDI externalities from national level FDI industry participation. In Nuevo León, this effect does not materialise. Furthermore, all three states are subject to negative intra-industry FDI externalities at the regional level. As for the importance of the determinants of FDI spillovers, Mexican firms in both Nuevo León and Jalisco enjoy additional positive FDI externalities in industries that are characterised

---

8   The variable FOR-national does vary across the regional industries. For a given regional industry, I calculate the level of foreign participation at the national level after subtracting the regional industry from the national aggregate.

Table 5.8 FDI externalities and interaction effects: Selected states

	Pooled sample	Estado de México	Jalisco	Nuevo León	Baja California	Chihuahua	Tamaulipas
Constant	3.16 (0.18)a	2.87 (0.54)a	3.36 (0.47)a	2.47 (0.40)a	3.34 (0.40)a	3.32 (0.51)a	2.24 (0.68)a
$k_{(m)}$	0.28 (0.03)a	0.33 (0.08)a	0.39 (0.06)a	0.30 (0.06)a	0.16 (0.06)b	0.18 (0.07)b	0.32 (0.08)a
$LQ_{(m)}$	0.12 (0.04)a	0.14 (0.09)	-0.09 (0.09)	0.21 (0.07)a	0.15 (011)	0.05 (0.12)	0.19 (0.17)
Herfi	0.09 (0.03)a	0.15 (0.06)b	0.11 (0.10)	0.13 (0.067)c	0.036 (0.06)	0.07 (0.10)	0.13 (0.14)
Gini	0.07 (0.045)c	0.45 (0.15)a	-0.15 (0.07)b	0.01 (0.08)	-0.09 (0.08)	-0.03 (0.11)	0.11 (0.10)
FOR-national	0.30 (0.08)a	0.31 (0.14)a	0.75 (0.24)a	-0.16 (0.19)	0.11 (0.20)	0.38 (0.20)c	-0.006 (0.27)
FOR-regional	-0.08 (0.02)a	-0.10 (0.05)a	-0.17 (0.04)a	-0.15 (0.06)b	-0.24 (0.12)b	-0.35 (0.11)a	-0.52 (0.28)c
GAP*FOR	0.02 (0.005)a	0.004 (0.008)	0.03 (0.007)a	0.13 (0.03)a	0.19 (0.07)b	0.03 (0.03)	0.12 (0.14)

Table 5.8 continued FDI externalities and interaction effects: Selected states

	Pooled sample	Estado de México	Jalisco	Nuevo León	Baja California	Chihuahua	Tamaulipas
GINI*FOR	-0.03 (0.03)	0.05 (0.07)	-0.06 (0.06)	0.02 (0.05)	-0.211 (0.17)	-0.16 (0.11)	0.12 (0.13)
TECH*FOR	0.07 (0.028)b	0.05 (0.16)	0.17 (0.09)c	0.04 (0.057)	0.12 (0.06)b	-0.05 (0.09)	0.11 (0.06)c
GINI*GAP*FOR	0.009 (0.004)b	0.04 (0.008)a	-0.013 (0.012)	0.09 (0.03)a	-0.04 (0.04)	0.001 (0.002)	-0.08 (0.09)
GINI*TECH*FOR	0.001 (0.01)	0.04 (0.03)	0.16 (0.05)a	0.009 (0.01)	-0.06 (0.05)	0.12 (0.04)a	0.11 (0.05)b
Adj. R²	0.49	0.53	0.75	0.74	0.39	0.52	0.64
F	29.20 (0.00)	13.37 (0.00)	115.20 (0.00)	18.08 (0.00)	4.46 (0.00)	15.30 (0.00)	32.13 (0.00)
N	457	110	55	64	71	57	53

Notes: Estimated standard errors in parentheses; robust to heteroscedasticity using Hubert/White/Sandwhich method. a, b, c, indicate significance levels at 1, 5 and 10%.

Estimated standard errors for pooled sample robust to clustering at the regional level. Pooled states regression includes state fixed effects.

In all estimations, the variable FOR-national is instrumented as in specification IV_2 in Table 5.5.

by a relative large technology gap. Furthermore, the positive effect of industry agglomeration on FDI spillovers, provided the level of technological differences is sufficiently large, is confirmed for all three states. In the case of Estado de México and Nuevo León, the positive effect of agglomeration is captured by the interaction variable between Gini, Gap and FOR, whereas for Jalisco the effect is indicated by the interaction variable between Gini, Tech and FOR.

The last three columns present the findings of FDI externality effects in the maquiladora states of Baja California, Chihuahua and Tamaulipas, states that experienced a rapid growth following the introduction of trade liberalisation. An important difference between the findings for these states and the findings for the other states is that the border states do not enjoy positive FDI externalities from FDI industry participation at the national level. This suggests that states with a shorter history of substantial involvement in manufacturing activity have not (yet) been able to benefit from the general industry presence of FDI firms in the country. As for intra-regional FDI effects, the findings indicate the presence of negative FDI externalities in all three states. Baja California and Tamaulipas both enjoy additional positive externalities in those industries with relative large technological differences between FDI and Mexican firms, in line with the findings for the other states. Furthermore, the estimations also identify the positive effect of industry agglomeration on FDI spillovers in two of the three border states, again with the qualification that the positive effect of industry agglomeration only materialises when the industry technology gap is sufficiently large.

6. Summary and Conclusions

Several previous studies on FDI effects in the Mexican economy have provided important evidence that FDI firms generate significant positive intra-industry productivity effects among Mexican manufacturing firms. The limitation of these findings is that they are based on the analysis of data for the 1970s and 1980s, time periods that are characterised by import substitution and government intervention. Also, findings on the effect of the technology gap on positive externalities are conflicting and there is very limited evidence on the effect of industry agglomeration on the occurrence and nature of FDI externalities. The analysis presented in this chapter addresses these limitations, by estimating FDI externalities for the mid 1990s and investigating explicitly whether technological differences between FDI and Mexican firms and industry agglomeration or geographical proximity are important for these effects.

As the data that I use is of a cross-sectional nature, the first important issue that I address is the problem that the variable measuring intra-industry foreign participation may be endogenous to the estimated regression model, a problem which represents the core of the criticism of cross-sectional studies of FDI externalities. I introduce a new instrument for this endogenous variable, in the form of the general variation of FDI intensity across manufacturing industries.

With this instrument, I obtain unbiased estimations of FDI spillovers in Mexican manufacturing industries. Importantly, the comparison between OLS and IV estimations shows that the OLS estimation underestimates the occurrence of positive FDI externalities. This is in strong contrast to the common criticism of cross-sectional estimates of FDI spillovers, which holds that these estimates are biased upwards, caused by a tendency among FDI firms to gravitate towards high productivity industries. In contrast, in the present analysis I find that FDI firms gravitate towards low productivity, labour intensive, manufacturing industries. Controlling for this tendency, the IV estimations produce robust evidence of substantial positive intra-industry FDI externalities.

Next, the analysis looks at the effects of determinants of FDI externalities in the form of the technology gap and industry agglomeration. A comparison of the estimated effect of intra-industry foreign participation between industries with a small or large technology gap indicates that positive FDI spillovers only materialise in the latter type of industry. Furthermore, estimations for both national level manufacturing industries and industries in selected individual states produce an estimated positive effect of an interaction variable between industry foreign participation and the industry technology gap, in support of the notion that large technological differences between FDI and Mexican firms enhance positive externalities. This finding is in strong contrast to the interpretation of the technology gap as direct inverse indicator of the level of absorptive capacity of domestic firms. Instead, the positive effect of the technology gap on positive FDI effects indicates the importance of a sufficiently large scope of potential externality effects. This provides incentives to Mexican firms to make externality facilitating investments that enhance their level of absorptive capacity. Also, it is likely that negative pecuniary externalities from competition are absent or less severe when there are large technological differences between FDI and Mexican firms.

Although the hypothesis that agglomeration will affect FDI spillovers is straightforward, given the large extent of similarity between channels of FDI spillovers and mechanisms of agglomeration economies, as well as the general positive effect of geographical proximity on knowledge spillovers, there is little direct empirical evidence on this effect. The findings presented in this chapter are in support of the hypothesis that agglomeration affects FDI externalities. Comparing the estimated effect of industry foreign participation between industries that are agglomerated to a high or a low extent shows that FDI spillovers materialise in the first type of industry, indicating the positive effect of agglomeration on these FDI effects. Having said so, the findings for the selected Mexican states indicate that the effect appears to be more complex: whereas the effect of intra-industry foreign participation at the national level is positive, at the regional level intra-industry foreign participation generates negative externalities. The estimated negative effect at the regional level may indicate that the negative competition effect has a regional dimension, or that the increase in prices of regional inputs that follows from the presence of FDI firms has a negative effect on the efficiency level of Mexican firms. In addition to this, the estimated effect of interaction variables between foreign participation and industry agglomeration further support the notion that agglomeration can foster positive FDI spillovers. The qualification to this finding is that the positive effect of agglomeration on FDI externalities appears to apply in particular to those industries that are also characterised by a sufficiently large technology gap between FDI and Mexican firms. This suggests that agglomeration enhances positive FDI externalities when there is a large potential for these effects to materialise and when negative effects from competition between Mexican and FDI firms are less likely to be important.

Chapter 6

FDI in Mexican Regions: Identifying the Industry and Geographical Dimensions of FDI Externalities

1. Introduction

Recent applied economics research on FDI productivity effects is developing new strategies to improve the identification of the full range of externality effects that can be linked to the presence and operations of FDI firms and to establish a better understanding of the conditions under which these effects materialise. The previous chapter linked to this latter research angle, by investigating whether and how the technology gap between FDI and Mexican firms and the industry level of agglomeration are influencing intra-industry FDI spillovers.

In the present chapter, I continue the empirical analysis of FDI externalities, focusing on the full identification of the industry and geographical dimensions of FDI effects. The industry dimension refers to the important recent recognition that FDI effects may arise both within and between industries. The geographical dimensions can be addressed in several ways. The importance of industry agglomeration can be assessed by estimating the effect of agglomeration on FDI spillovers, and by distinguishing between national and regional FDI effects. As the findings in the previous chapter show, both these features are important for intra-industry FDI externalities. In the present chapter, I assess whether this is also the case for inter-industry externalities. In addition, I extend the analysis of geographical dimensions by estimating for the presence of intra- and inter-regional FDI effects and by assessing whether regional characteristics influence these spatial spillovers.

The chapter is constructed as follows. In section 2 I briefly discuss the industry and geographical dimensions of FDI externalities. In section 3 I discuss the datasets that I use in this chapter. One dataset consists of detailed industry observations for selected Mexican states, as introduced in the previous chapter. The other dataset contains more aggregated industry observations for all 32 Mexican states. Section 4 presents the empirical findings. With the dataset for all 32 states, I estimate intra-regional intra- and inter-industry FDI spillovers. The dataset for selected Mexican states allows me to estimate for externality effects among Mexican supplier firms, customer firms and competitors, both at the national and the regional level. Finally, with the dataset for all Mexican states I estimate for the presence of intra- and inter-regional FDI externalities and for the influence of regional characteristics on these spatial FDI effects. Section 5 summarises and concludes.

2. The Industry and Geographical Dimensions of FDI Externalities[1]

The notion that the full industry dimension of FDI spillovers needs to be controlled for in estimations of FDI effects follows from the recognition that FDI effects may occur within and between industries. The substantial degree of similarity between the channels of FDI spillovers and mechanisms underlying agglomeration economies suggests that, like agglomeration economies, FDI spillovers may be of an intra- and inter-industry nature (Jordaan 2004a, 2008b). Also, one could argue that FDI effects between industries are more likely to occur, as FDI firms, in their attempts to minimise positive spillover effects among their competitors, focus their efforts on minimising intra-industry effects (Kugler 2006). Furthermore, characteristics of input markets (Lall 1980) and the desire of FDI firms to improve the performance of local suppliers in host economies often lead to the provision of several types of support to these suppliers, resulting in the occurrence of positive FDI externalities between industries (UNCTAD 2001, Blomström and Kokko 1998).

As indicated in the survey in Chapter 2, there are several studies that produce evidence of positive inter-industry FDI effects, in particular between FDI and local suppliers (e.g. Liu 2008 for China, Svejnar 2007 for a large selection of eastern European countries). However, other studies find that these inter-firm linkages may also generate negative FDI externalities (e.g. Yudaeva et al. 2003 for Russia, Javorcik et al. 2004 for Romania and Driffield et al. 2004 for the UK). In extension of these findings, several other studies have combined the search for inter-industry FDI externalities with the incorporation of a geographical dimension of these effects, by estimating the effect of inter-industry foreign participation at the regional level within host economies. A well known example is Blalock and Gertler (2008), who present evidence that inter-industry regional FDI participation is positively associated with productivity of Indonesian manufacturing plants. Other examples of positive findings of this type of FDI variable include Girma and Wakelin (2007) for the UK and Halpern et al. (2007) for Hungary. Again, however, the evidence is heterogeneous in nature, as other studies find no significant effect of this type of regional FDI participation (e.g. Aitken and Harrison 1999 for Venezuela, Haskel et al. 2007 for the UK). Also, as discussed in Chapter 2, caution is required in interpreting the findings from several of these studies. One issue is that most studies fail to control for the possibility that there may be separate FDI effects at the national and regional level. As indicated by the findings in the previous chapter, this may lead estimations to miss out on identifying part of the FDI effects. In relation to this, the use of a variable that lumps together regional intra- and inter-industry FDI participation introduces the possibility that the estimated effect of this composite regional FDI variable is affected by aggregation bias. Findings from

----

1   In this section, I briefly discuss the industry and geographical dimensions of FDI externalities, using key papers to indicate the main issues. Chapter 2 contains a more extensive discussion of these issues.

other studies may be influenced by omitted variable bias, caused by estimating only for the effect of regional intra- or inter-industry FDI participation.

Geographical Dimensions of FDI Externalities

There are several interpretations of the geographical dimensions of FDI externalities. One interpretation relates to the notion that industry agglomeration is likely to influence FDI effects. The few studies that estimate the effect of industry agglomeration or geographical proximity between firms produce evidence supporting the notion that agglomeration enhances positive FDI spillovers (e.g. De Propris and Driffield 2006, Barrios et al. 2006). Also, the findings presented in the previous chapter confirm this positive effect in Mexican manufacturing industries for intra-industry FDI externalities.

Another approach is to control for the effect from foreign participation in all regions in a host economy. An example of this is Smarzynska (2002), who estimates FDI spillover effects within and between regions in Lithuania, also distinguishing between intra- and inter-industry foreign participation. The empirical findings confirm the presence of inter-industry FDI effects through backward linkages between FDI and domestic suppliers, both at the intra- and inter-regional level. Driffield (2006) presents findings from an industry-level study of FDI externalities in UK regions. In this study, Driffield similarly distinguishes between intra- and inter-industry foreign participation both within and between regions. The findings indicate the presence of positive intra- and inter-industry FDI effects, both at the intra-regional level. There appear to be no FDI effects that arise between UK regions. Having said so, in a related study, Driffield et al. (2004) present evidence of a variety of intra- and inter-regional FDI effects. In particular, the estimations identify positive intra- and inter-regional FDI effects through forward linkages and negative FDI externalities through backward linkages between UK regions. Finally, Girma and Wakelin (2007) and Driffield and Girma (2003) estimate FDI effects among UK manufacturing plants but find no evidence of inter-regional effects.

As is the case with the evidence on the industry dimension of FDI spillovers, the empirical findings on intra- and inter-regional FDI effects are characterised by a substantial level of heterogeneity. Again, the evidence is sometimes difficult to interpret, as studies do not always control for the full industry and geographical dimensions of FDI effects. Furthermore, the majority of estimations of inter-regional FDI effects are based on ad-hoc specifications of the relation between geographical distance and spatial FDI effects. Given the impossibility to determine this relationship *a priori*, the best strategy to identify spatial spillovers empirically is to experiment with several distance decay parameters (Anselin 1988, Bode 2004, Jordaan 2008b). The need to do so is further underlined by the fact that FDI externalities can be caused by a variety of mechanisms, which may be affected by geographical distance differently. Finally, it appears that most studies have not given full consideration to the use of an appropriate indicator of the magnitude of potential spatial FDI effects. In particular, the reliance on spatially weighted variables that are simply the indicators of intra- and inter-industry regional foreign participation creates the possibility that the scale of these spatial FDI effects is measured incorrectly.

3. Data, Production Function and Variables

For the empirical analysis of this chapter, I use two datasets. One dataset contains unpublished industry level data for selected Mexican states, as described in the previous chapter. The other dataset also consists of unpublished data from the 1994 Mexican manufacturing census, containing more aggregated 2-digit industry observations for all 32 Mexican states, again distinguishing between Mexican and foreign-owned industry aggregates. The industries are; (31) Food, drinks and tobacco, (32) Textiles, clothing and leather, (33) Wood and wood products, (34) Paper, paper products and printing, (35) Chemicals, rubber and plastic products, (36) Non-metallic minerals, (37) Basic metals, (38) Metal products and equipment and (39) Other manufacturing activities. The dataset contains value added, total assets at book value, number of employees, number of white collar employees and number of blue collar employees. After deleting industries with missing values for value added or number of employees, the dataset contains 175 regional industries with usable observations.

The nature of the dataset makes the standard Cobb-Douglas function the best suited to derive the regression model from (see Intrilligator et al. 1996). This production function can be depicted as:

(1) $Q = A * K^{\alpha} * L^{\beta}$;

where Q, K and L are production, capital and labour, A is the efficiency parameter and α and β are the output elasticities of capital and labour. Expressing this function in its intensive form gives a function that can be estimated with the data:

(2) $Q/L = A * (k)^{\alpha}$;   where $k = (K/L)$.

A potential shortcoming of equation (2) is that it is based on the assumption of constant returns to scale. To circumvent this assumption, I use an alternative derivation of the CD function that does allow for the presence of scale economies. Instead of assuming constant returns to scale $(\alpha+\beta=1)$, I assume that $(\alpha+\beta=\delta)$, where δ can be smaller, equal to or larger than 1. This alternative assumption gives:

(3) $Q = A * K^{\alpha} * L^{\beta}$;

where $\beta = (\delta-\alpha)$. Subtracting 1 from both sides gives $\beta -1 = (\delta-\alpha-1)$. Substitution into (3) gives:

(4) $Q = A * K^\alpha * L^{(\delta-\alpha-1)}$.   Dividing both sides by L gives the intensive function:

(5) $Q/L = A * (k)^\alpha * L^{(\delta-1)}$.

Taking logs and adding the industry (i) and regional (r) dimensions gives the equation that can be estimated directly with the dataset:

(6) $\ln(Q/L)_{(m)ir} = \ln A_{ir} + \alpha \ln (k)_{(m)ir} + (\delta-1) \ln L_{(m)ir} + \varepsilon_{ir}$;

where (m) stands for Mexican-owned industry share and ε is the error term of the estimation. Equation (6) contains one unknown parameter $(\delta-1)$ for which I use a proxy variable in the regression model.

Table 6.1 lists the variables that I use to estimate the regression model. Summary statistics are shown in Table 6.2. The dependent variable is measured as the ratio of value added over the number of manufacturing employees of a Mexican-owned industry share. The capital-labour ratio is measured as the ratio of total assets at book value over the number of manufacturing employees of Mexican firms. To capture scale economies, I need to define a proxy for the term $(\delta-1)$. I calculate this proxy using the concept of minimum efficient scale or MES (see e.g. Tybout and Westbrook 1995). This concept refers to the production volume at which long run production costs are minimised. I approximate MES production by the average production of the largest plant size in a regional industry. As proxy for the term $(\delta-1)$ I take the ratio of average production per plant in a regional industry over MES production.

Other factors that may affect labour productivity of Mexican firms are included in the efficiency parameter A. One variable that I include is human capital, measured as the ratio of white collar employees over blue collar employees of a Mexican-owned industry share. Second, I include a variable that controls for the presence of duality within regional industries. This duality refers to the situation that within Mexican manufacturing industries, traditional and modern segments co-exist with little to no inter-relations (Blomström 1989). Traditional segments consist of micro and small firms, using traditional technologies and operating at low productivity levels. Modern segments consist mainly of high productivity medium and large firms that use modern technologies. To control for the co-existence of these different segments within regional industries, I include the variable DUAL, which I measure as the ratio of the number of firms over the number of employees in a regional industry divided by the ratio of the number of firms over the number of employees in the national industry. I expect a negative effect of this variable, as a high score indicates that a regional industry has a relative large traditional segment.

Table 6.1   Definition of variables

Variable	Description	Definition
$Q/L_{(m)ir}$	Labour productivity Mexican owned share of industry	$$\frac{(valueadded)(m)ir}{(numberofemployees)(m)ir}$$
$(k)_{(m)ir}$	Capital-labour ratio Mexican owned industry share	$$\frac{(totalassetsatbookvalue)(m)ir}{(numberofemployees)(m)ir}$$
$LQ_{(m)ir}$	Human capital Mexican-owned industry share	$$\frac{(numberofwhitecollaremployees)(m)ir}{(numberofbluecollaremployees)(m)ir}$$
$(\delta\text{-}1)$	Scale economies parameter	$$\frac{(production/firms)ir}{MESir}$$ $$MESir = \frac{(production)ir}{(firms)ir} \quad \text{largest plant size}$$
$Dual_{ir}$	Indicator of participation of traditional production technologies	$$\frac{(numberoffirms/numberofemployees)ir}{(numberoffirms/numberofemployees)i - national}$$

Table 6.1 continued   Definition of variables

Variable	Description	Definition
Industrymix$_{ir}$	Predictor of effect regional industry mix (see text for elucidation)	$$\sum W \frac{(valueadded)i - national(4digit)}{(numberofemployees)i - national(4digit)}$$ $$W = \frac{(numnberofemployees)ir(4digit)}{(numberofemployees)r(2digit)}$$
Density$_{ir}$	Density of regional manufacturing sector	$$\sum U \frac{(employees)r - municipality}{(squarekilometers)r - municipality}$$ $$U = \frac{(employees)r - municipality}{(employees)r}$$
FOR$_{ir}$	Intra-industry foreign participation	$$\frac{(numberofemployees)(f)ir}{(numberofemployees)ir}$$
FIR$_{ir}$	Inter-industry foreign participation	$$\left(\frac{(numberofemployees)(f)r - (numberofemployees)(f)ir}{(numberomemployees)r - (numberofemployees)ir} \right)$$

Note: (m) = Mexican-owned industry share; (f) = Foreign-owned industry share; i, r = industry, region.

Source: Q/L$_{(m)ir}$, (k)$_{(m)ir}$, LQ$_{(m)ir}$, FOR$_{ir}$, FIR$_{ir}$ based on unpublished data provided by INEGI, other variables calculated with data from the 1994 economic census.

Table 6.2   Summary statistics

	Mean	St. dev.	Minimum	Maximum
$Y/L_{(m)ir}$	3.37	0.867	0.235	6.848
$(k)_{(m)ir}$	3.81	1.113	1.006	7.17
$LQ_{(m)ir}$	-1.28	0.623	-3.20	0.66
$(\delta-1)L_{(m)ir}$	1.43	2.01	0.06	12.86
$Dual_{ir}$	0.301	0.89	-1.77	4.28
$Industrymix_{ir}$	3.90	0.60	2.87	5.61
$Density_{ir}$	3.92	1.64	3.34	8.11
For_{ir}	-2.56	1.83	-9.28	0
Fir_{ir}	-2.54	1.44	-8.21	-0.35

Note: All variables are in logs; (m) = Mexican-owned industry share; i, r = industry, region.

I also need to control for the possibility that differences in productivity between regional industries are caused by differences in the industry mix of these industries (see Rigby and Essletzbichtler 1997, 2000, Jordaan 2008b), rather than by differences in for instance human capital or capital intensity. To capture the industry mix effect, I calculate a predictor variable as follows. I calculate the ratio of value added over the number of manufacturing employees for national 4-digit industries. Then, for a given 2-digit regional industry, I calculate the predictor of the industry mix effect by weighting the national 4-digit productivity indicators with 4-digit regional industry employment shares of the 2-digit regional industry.

Next, I include a variable to control for the effect of agglomeration economies. For similar reasons as discussed in Chapter 4, I follow Ciccone and Hall (1996) in using density instead of scale variables to capture these place-based externalities. Again, as density calculated with state level data may not produce an accurate indicator of density of activity within a state, I use municipality level data. I calculate the level of density of the manufacturing sector for all municipalities of a state, dividing the number of manufacturing employees by the size of the municipalities. I then aggregate the municipality density scores, using the municipalities' share in state level manufacturing employment as weight.

Finally, I add variables to the regression model that control for externality effects from FDI. I measure intra-industry foreign participation as the employment share of FDI firms in a regional industry. Furthermore, I add a variable capturing inter-industry foreign participation, measured, for a given industry, as the employment share of FDI in total state manufacturing employment minus employment of the given industry.

4. Intra- and Inter-industry Externalities from FDI: Empirical Findings

Intra-regional Externalities

The base line regression model that I estimate with the regional dataset to identify FDI externalities is (dropping ln, i, r notation):

(6.A) $Q/L_{(m)} = \beta0 + \beta1(k)_{(m)} + \beta2\ LQ_{(m)} + \beta3\ Dual + \beta4\ (\delta-1)L_{(m)}$

$\beta5\ Industrymix + \beta6\ Density + \beta FOR + \beta FIR + \varepsilon.$

The estimator is OLS. The main findings from estimating several specifications of regression model (6.A) are presented in Table 6.3. The first column with findings contains the results from estimating the regression model without the controls for foreign participation. The positive effect of the capital-labour ratio indicates that labour productivity is higher in capital intensive industries. The positive effect of the industry mix variable indicates that part of the difference in productivity between regional industries is not caused by structural differences between these industries, but rather by differences in the mix of industrial activities of these regional industries. The negative coefficient of the duality variable shows the expected negative effect of the size of the traditional segment of a regional industry on industry productivity. The level of density of the manufacturing sector has a positive effect, pointing to the presence of positive agglomeration economies. An additional regression where I substitute the density level of individual regional industries for density of the regional manufacturing sector (results not reported) produces an estimated insignificant effect of the density variable, suggesting that agglomeration economies are of a multi-industry nature.[2]

The next column shows the results from adding the intra-industry foreign participation variable. The estimated effect of this variable is negative, suggesting the presence of negative intra-industry intra-regional FDI externalities. This finding is in line with the findings from Chapters 4 and 5. To assess whether there is a problem with endogeneity, and in the absence of a suitable instrument, I follow the second best solution and use a lagged value for the foreign participation variable, calculated with unpublished data provided by INEGI from the 1989 manufacturing census (containing observations for 1988). As the findings in column three indicate, the estimated effect of the lagged foreign participation variable persists to be negative, supporting the impression that there are negative intra-industry FDI externalities.[3]

---

2   The estimated effect of labour quality is insignificant. Estimations without industry and regional fixed effects produce a positive significant effect of labour quality, suggesting that the effect of labour quality is captured by the regional and/or industry dummies.

3   I did experiment with converting the instrument of Chapter 5 to the more aggregated industry level; unfortunately, at the more aggregated industry level the instrument does not function satisfactory.

Table 6.3   Intra-regional FDI externalities

	1	2	3	4	5	6
Constant	-0.06 (0.34)	-0.15 (0.34)	-0.33 (0.42)	-0.11 (0.34)	-0.48 (0.44)	-0.42 (0.42)
$K_{(m)}$	0.37 (0.04)a	0.38 (0.04)a	0.38 (0.05)a	0.37 (0.04)a	0.36 (0.05)a	0.36 (0.05)a
$LQ_{(m)}$	0.03 (0.06)	0.06 (0.06)	-0.04 (0.08)	0.05 (0.06)	-0.06 (0.08)	-0.04 (0.09)
Dual	-0.15 (0.04)a	-0.19 (0.04)a	-0.13 (0.06)b	-0.15 (0.04)a	-0.13 (0.06)b	-0.14 (0.05)a
$(\delta-1)L$	-0.02 (0.02)	-0.01 (0.02)	0.01 (0.02)	0.01 (0.02)	0.02 (0.02)	0.02 (0.02)
Industrymix	0.50 (0.08)a	0.51 (0.08)a	0.51 (0.10)a	0.53 (0.08)a	0.57 (0.11)a	0.55 (0.10)a
Density	0.07 (0.02)a	0.06 (0.02)a	0.08 (0.02)a	0.06 (0.02)a	0.07 (0.02)a	0.06 (0.02)a
FOR	–	-0.04 (0.02)b	-0.04 (0.02)b	-0.04 (0.017)a	-0.05 (0.02)a	-0.05 (0.02)a
FIR	–	–	–	0.06 (0.03)b	0.07 (0.03)b	0.07 (0.02)a
Technologygap	–	–	–	–	–	0.11 (0.03)a
State and industry effects	Yes (0.00)	Yes (0.00)	Yes (0.00)	Yes (0.00)	Yes (0.00)	Yes (0.00)
Adj. R^2	0.79	0.79	0.74	0.79	0.75	0.79
N	175	175	138	175	138	175

Notes: (m) – Mexican-owned industry share.

Estimated standard errors shown in parentheses, robust to heteroscedasticity and clustering at the state level. a, b and c indicate significance level of 1, 5 and 10%. State and industry effects statistic is F test whether state and industry dummies have coefficients equal to 0.

Regression of column 3 uses lagged FOR, regression of column 5 uses lagged FOR and FIR.

The estimated effect of scale economies is insignificant in all the estimations. As a check, I have estimated alternative regression models in which the parameter $(\delta-1)$ ranges from 0.8 (mild decreasing returns to scale) to 1.2 (mild increasing returns to scale). The estimated effect of the other control variables is stable in all these additional regressions, indicating that any problems with capturing the presence of increasing returns to scale do not appear to influence the identification of the effects of the other variables.

Column four presents the findings from adding inter-industry intra-regional foreign participation to the regression model. The estimated effect of intra-regional intra-industry foreign participation is not affected by the inclusion of this variable and remains negative. The estimated effect of the inter-industry FDI variable is positive, suggesting the presence of positive inter-industry FDI externalities. Again, I re-estimate the regression model, using lagged values for both FDI variables. As the results in column five indicate, the estimated effect of the two foreign participation variables is the same with the lagged values, suggesting that the estimated effect of the FDI variables is not affected by endogeneity.

Finally, I address the effect of the technology gap on FDI effects. Using data for 1988, I follow Blomström and Wolff (1994) by measuring the technology gap between FDI and Mexican firms as the ratio of value added per manufacturing employee in FDI firms over value added per manufacturing employee in Mexican firms. As the findings in the last column show, the estimated effect of the technology gap variable is positive, indicating a positive effect of large technological differences between FDI and Mexican firms on productivity of these Mexican firms. Again, this finding is in support of the hypothesis that large technological differences are conducive rather than detrimental to FDI spillover effects, which is in contrast to the interpretation of the technology gap as direct inverse indicator of the level of absorptive capacity of Mexican firms. Instead, it points at the importance of a sufficient scale of potential FDI externalities, incentives among Mexican firms to make externality-facilitating investments to improve their level of absorptive capacity and the likely absence of negative effects from competition between FDI and Mexican firms.

Intra-regional FDI Externalities: Evidence from Selected States

The dataset with pooled industry observations for the selected states allows me to obtain additional evidence on intra- and inter-industry FDI effects. With this dataset, I estimate the following regression model, which is an augmented version of the regression model of Chapter 5:

(6.B)
$$Q/L_{(m)} = \beta 0 + \beta 1(k)_{(m)} + \beta 2\ LQ_{(m)} + \beta 3\ (\delta\text{-}1)L_{(m)} + \beta 4\ HERFI + \beta 5\ GINI +$$

$$\beta 6\ FORnational + \beta 7\ FORregional + \beta 8\ BWnational + \beta 9\ BWregional$$

$$+ \beta 10\ FWnational + \beta 11\ FWregional + \beta 12\ FOR*GAP$$

$$+ \beta 13\ FOR*GINI + \beta 14\ BW*GAP + \beta 14\ BW*GINI + \beta 15\ FW*GAP$$

$$+ \beta 16\ FW*GINI + \varepsilon.$$

This augmented regression model contains additional variables capturing inter-industry FDI effects to Mexican suppliers and client firms. Following Blalock and Gertler (2008), I calculate the level of FDI participation in industries to which Mexican firms supply inputs to – backward linkages – as follows:

$$BW_{ir} = \sum_{k} c_{ir}k\ FOR_{kr}$$

where $c_{ir}k$ is the proportion of sector i output that is consumed by sector k in region r. In a similar fashion, I calculate the level of foreign participation in industries that supply inputs to Mexican firms – forward linkages – as:

$$FW_{ir} \sum_{k} d_{ir}k \, FOR_{kr}$$

where $d_{ir}k$ is the proportion of sector i that is supplied by sector k in region r. For the calculation of both variables, I use an input-output Table for 1993 provided by INEGI. I calculate these forward and backward linkages variables both at the regional and the national level.[4] All interaction variables between the foreign participation variables and Gap or Gini are calculated with the regional foreign participation variables.

The main findings from estimating regression model (6.B) are presented in Table 6.4. The first column with results contains the findings from estimating the regression model without any interaction variables. There is a positive effect of regional backward linkages, suggesting the presence of positive externalities between FDI firms and Mexican suppliers. The importance of geographical proximity for these effects is indicated by the fact that there are no externalities through backward linkages at the national level. In contrast, the estimated effect of the variable capturing FDI forward linkages at the national level is mildly significant, whereas the variable of forward linkages at the regional level is not significantly associated with productivity of Mexican firms. As for intra-industry effects, the findings are in line with the previous chapter, indicating negative effects at the regional level and positive effects at the national level from intra-industry foreign participation.

In extension of these findings, the second column with results presents the findings from estimating regression model (6.B) with the interaction variables between the foreign participation variables and the technology gap (Gap) and industry agglomeration (Gini). The findings present a good impression of the variety of externality effects that are linked to the presence of FDI firms in the manufacturing industries of the selected states. Importantly, except for the effect of forward linkages at the national level which turns insignificant, the estimated effects of the foreign participation variables from the regression model without the interaction variables are maintained. As for the effect of the interaction variables, large technological differences between FDI and Mexican firms again have a positive effect. Intra-industry foreign participation at the regional level generates positive FDI spillovers in industries with a relative large technology gap. Also, a similar positive effect from the technology gap applies to inter-industry FDI

---

4   The variables of intra-industry and inter-industry foreign participation at the national level do vary across the regional industries, as I calculate these variables after having subtracted regional industries from the national aggregate.

Table 6.4   Intra- and inter-industry FDI externalities in selected states:
              Pooled sample

	1	2
Constant	4.10 (0.48)a	3.76 (0.22)a
$k_{(m)}$	0.23 (0.05)a	0.20 (0.04)a
$LQ_{(m)}$	0.11 (0.04)b	0.11 (0.04)b
$(\delta-1)L_{(m)}$	0.02 (0.03)	0.02 (0.02)
HERFI	0.22 (0.03)a	0.25 (0.05)a
GINI	0.17 (0.04)a	0.06 (0.08)
FORregional	-0.06 (0.02)a	-0.08 (0.03)b
FORnational	0.17 (0.07)b	0.21 (0.04)a
BWregional	0.09 (0.04)b	0.08 (0.03)b
BWnational	-1.80 (1.20)	0.009 (0.04)
FWregional	0.005 (0.01)	-0.006 (0.007)
FWnational	0.02 (0.012)c	0.006 (0.009)
FOR*GAP	−	0.011 (0.005)b
FOR*GINI	−	0.008 (0.03)
BW*GAP	−	0.08 (0.02)a
BW*GINI	−	-0.06 (0.03)b
FW*GAP	−	0.004 (0.002)b
FW*GINI	−	-0.03 (0.02)
State and industry effects	10.35 (0.00)	94.09 (0.00)
Adj. R^2	0.50	0.55
F	16.15 (0.00)	24.52 (0.00)
N	460	460

Notes: Estimated standard errors in parentheses, robust to heteroscedasticity and clustering at the state level.

a and b indicate significance level at 1 and 5%; States are Estado de México, Jalisco, Nuevo León, Baja California, Coahuila, Chihuahua, Sonora and Tamaulipas. Interaction variables between GAP or GINI with foreign participation variables are calculated with FORregional, BWregional and FWregional; FORnational is instrumented following the IV_2 specification of Table 5.5 in Chapter 5.

externalities at the regional level, both through backward linkages and forward linkages. In addition to these findings, the estimated effect of the interaction variable between industry agglomeration and FDI backward linkages is negative, which suggests that for this type of FDI effect industry agglomeration fosters negative externalities among Mexican suppliers.

Although there is some corroborating evidence for the occurrence of negative FDI externalities through backward linkages (e.g. Yudaeva 2003, Javorcik et al. 2004, Driffield et al. 2004), explanations for this effect are scant, especially in combination with the feature that industry agglomeration fosters such a negative effect. One possible explanation is that the presence of an agglomeration fosters fierce competition between Mexican firms that operate as supplier to FDI firms. This may force them to offer levels of concession (e.g. lower prices, shorter delivery times) that outweigh any support provided by FDI firms, resulting in the net creation of negative externalities. In relation to this, one could also argue that the scale of support provided by FDI firms may be lower in an agglomeration, given the large quantity of suppliers, making the occurrence of negative externalities more likely.

One important issue that is not captured by the estimations of regression model (6.B) is that the FDI effects may differ across the selected states. It is likely that such regional heterogeneity exists, given the substantial changes that these states experienced following the introduction of trade liberalisation in the late 1980s. To see whether this is the case, I re-estimate the regression model, adding interaction terms between state dummies and the interaction variables between the foreign participation variables and the variables Gap and Gini. A summary of the findings from these additional estimations is presented in Table 6.5.

It is clear from the findings that there are indeed additional FDI effects that differ between the states. Mexican firms in Baja California, one of the more important border states in terms of participation in Mexico's manufacturing sector, enjoy additional FDI effects, as both the technology gap and industry agglomeration appear to enhance positive FDI effects. This positive effect applies to both intra-industry externalities and externalities through backward linkages. Coahuila and Nuevo León enjoy additional intra-industry externalities in industries with a large technology gap. Furthermore, Nuevo León and Tamaulipas benefit from additional positive externalities through backward linkages, again in industries with relative large technological differences between FDI and Mexican firms. The other two border states Chihuahua and Sonora do not appear to enjoy additional FDI externalities.

In strong contrast to these findings for the border states, the findings show that the agglomeration Mexico City and also the state Jalisco, which both were important during the period of import substitution, are subject to additional negative externalities. Most of these additional negative effects appear to be associated with the effect of industry agglomeration on negative intra-industry externalities as well as on effects through backward and forward linkages. As discussed earlier, the negative effect of agglomeration can be explained by a decrease in support

Table 6.5   Regional variation in externality impact of FDI

Region	FOR*GAP	FOR*GINI	BW*GAP	BW*GINI	FW*GAP	FW*GINI
Baja California	0.05 (0.014)a	0.10 (0.05)b	0.05 (0.02)b	0.05 (0.02)a	–	–
Coahuila	0.06 (0.025)b	–	–	–	–	–
Chihuahua	–	–	–	–	–	–
Jalisco	–	-0.06 (0.02)a	–	–	-0.03 (0.003)a	-0.04 (0.007)a
Estado de México	–	-0.02 (0.009)b	-0.01 (0.003)a	-0.06 (0.027)c	-0.04 (0.03)a	-0.06 (0.01)a
Nuevo León	0.05 (0.015)a	–	0.03 (0.013)a	–	–	–
Sonora	–	–	–	–	–	–
Tamaulipas	–	–	0.04 (0.015)a	–	–	–

Note: The findings are from estimating regression model (6.B), specification column (2) in Table 6.4, adding interaction terms between state dummies and interaction variables between FOR or FIR with GAP or GINI. a, b and c indicate significance levels at 1, 5% and 10%.

by FDI firms and an increase in concessions offered by Mexican suppliers, as a result of the high level of competition between the suppliers that is likely to occur in an agglomeration. This negative effect may be further strengthened by the fact that Mexico City and Jalisco have come under increased competition from the border states in the last two decades. This may have made FDI firms even less willing to engage in supporting relations, as the opening of the market has made it easier to source from international suppliers, thereby only increasing the level of competitive pressure on suppliers in Mexico City and Jalisco. Similarly, the negative effect on local client firms through forward linkages may be caused by a decreased willingness among FDI firms to offer support to these client firms, as the opening up of the market has also facilitated foreign-owned firms to engage in international trade. In combination, it is certainly interesting to see that Mexico City in particular has been subject to additional negative externalities, a situation which is very likely to have reinforced the process of relative decline that this agglomeration of economic activity has undergone in the last two decades.

Inter-regional Effects from FDI

In order to obtain improved indications of the full range of externality effects that can be associated with the presence and operations of FDI firms, the next step of the analysis is to incorporate spatial effects into the estimation. For the estimation of spatial FDI effects, I use the regional database with all the 32 Mexican states.

To capture inter-regional FDI effects, I need to adapt regression model (6.A) by adding variables that capture both the scope of potential inter-regional effects and the relation between spatial externalities and geographical distance. Doing so, the model becomes:

$$(6.C) \quad Q/L_{(m)} = \beta 0 + \beta_x X + \beta 6 \text{ FOR} + \beta 7 \text{ FIR} + \beta 8 \sum_{r'1;r'\neq r}^{r} W \text{ FORinterregional} +$$

$$\beta 9 \sum_{r'1;r'\neq r} W \text{ FIRirinterregional} + \varepsilon;$$

where $X = ((k)_{(m)}, (\delta\text{-}1)L_{(m)}, \text{Dual}, \text{Industrymix}, \text{Density})$.

W is a distance matrix containing spatial weights w_{rj}, capturing the relation between geographical distance and FDI externalities between regions r and j. The variables FORinterregional and FIRinterregional capture the size of potential inter-regional FDI effects.

As discussed in Chapter 4, it is not possible to determine the relation between geographical space and inter-regional effects *a priori* (see Anselin 1988, Bode 2004). In the case of estimating spatial FDI externalities, the need to experiment with several specifications of the distance matrix W is further indicated by the fact that there is a variety of mechanisms that may underlie these productivity effects, mechanisms that may be affected by geographical distance differently. Therefore, using only one specification of the distance matrix W runs the risk of failing to identify accurately the full range of spatial FDI effects.

I experiment with 10 different specifications of the distance matrix. I start from the assumption that geographical distance has no delimiting effect on spatial FDI spillovers. In other words, FDI in a given state may affect productivity of Mexican firms in all other states equally, irrespective of the fact that some states are located closer to the state than other states. In terms of the distance matrix, all w_{rj}'s are set to 1.

Second, I use the contiguity assumption. Under this assumption, spatial effects can only occur between states that share a border. This specification is based on the assumption that FDI effects have a spatial cut-off point after which they seize to materialise. In the empirical estimation, I try out first and second order contiguity. First order contiguity only considers potential spatial FDI spillovers between states that share a border, second order contiguity also includes states that share a border with the first order contiguity states. In terms of the distance matrix, using e.g. the first order contiguity assumption means that all w_{rj}'s take the value of 1 when states share a border and 0 otherwise. An alternative way to capture the existence of a spatial cut-off point in the process of spatial FDI effects is the nearest neighbour assumption. For instance, the k=2 assumption entails that spatial FDI effects can occur between a state and its two geographically closest neighbour states. The w_{rj}'s take the value of 1 for these states and 0 otherwise. In estimating the regression model, I try out k=2, k=3 and k=4.

Finally, I estimate regression models that are based on the assumption that spatial FDI effects are negatively related to geographical distance in a continuous fashion. This gravity-like specification entails that FDI in a given state can potentially affect Mexican firms in all other states, but the externality effects are assumed to be decreasing with increasing distance between the states. An often used specification is to define the w_{rj}'s as the inverse distance between regions (Anselin 1988, Adserá 2000). In the empirical analysis, I experiment with four specifications that give increasing weight to the distance decay effect, letting the spatial weighting variable increase from (1/distance) to (1/(distance)⁴). I measure distance between the states as the number of kilometres between state capital cities, based on the 'as the crow flies' assumption.

As for measuring the scope of potential inter-regional FDI effects, the common strategy in applied research on spatial FDI externalities is to simply use the share of FDI in a regional industry and in total regional manufacturing employment as indicators of the magnitude of potential inter-regional effects. As I discussed in Chapter 2, it seems likely that these variables do not provide accurate indicators of the magnitude of these potential spatial FDI externalities. For instance, in the case where a few FDI firms have a substantial share in employment of a small regional industry, the use of the share of FDI in regional industry employment would suggest that the potential of spatial intra-industry FDI effects from this regional industry is large, which is clearly not in line with the small absolute scale of FDI in this state. Similarly, using the share of FDI employment in a case where a substantial number of FDI firms operate in a large regional industry would lead to indicate that the scope of potential spatial FDI effects from this industry is relatively small, which is in contrast to the large scale of FDI activity in this industry.

To correct for this, I calculate an alternative indicator for the magnitude of potential spatial intra-industry FDI spillovers as follows:

$$FORinterregional_{ir} = \frac{(numberofemployees)_{ir}^{f}}{(numberofemployees)_{ir}} \times \frac{(numberofemployees)_{ir}^{f}}{\sum_{r=1}^{32}(numberofemployees)_{r}^{f}};$$

where f indicates the foreign-owned share of an industry and i and r represent the industry and regional dimensions of the data. The difference between this alternative indicator and the commonly used one is that the level of intra-industry foreign participation (the first term) is weighted by it share in total FDI industry employment in Mexico. In a similar fashion, I can also calculate an alternative indicator of the magnitude of potential spatial inter-industry FDI effects:

$$FIRinterregional_{ir} = \frac{(numberofemployees)_{r}^{f} - (numberofemployees)_{ir}^{f}}{(numberofemployees)r - (numberofemployees)ir} \times$$

$$\frac{(numberofemployees)_{r}^{f} - (numberofemployees)_{ir}^{f}}{\sum_{r=1}^{32}(numberofemployees)_{r}^{f}}$$

Again, this alternative indicator weights the traditional indicator of the magnitude of potential spatial inter-regional inter-industry FDI effects by its relative size with respect to total FDI employment in the Mexican manufacturing sector.

Empirical Findings on Inter-regional FDI Externalities

Table 6.6 presents the findings from estimating regression model (6.C) with the various specifications of the distance matrix and with the two different types of indicator of the magnitude of potential inter-regional FDI externalities. The first column with results contains the findings from the regression model based on the assumption that geographical distance does not affect inter-regional FDI effects. The results suggest the absence of spatial effects, as the variables of intra- and inter-industry inter-regional FDI participation carry insignificant coefficients. This partly corroborates the findings from Table 6.4, as those findings indicate that there are no national level inter-industry effects through forward and backward linkages among the selected states. Having said so, the findings for the selected states do identify a positive effect from intra-industry foreign participation at the national level. Most likely, this difference in findings can be explained by the fact that the estimations of FDI effects in the selected states concern estimations for those states that are most affected by FDI firms. The significant non-spatially weighted effect from overall intra-industry foreign participation that exists in these states may not materialise when estimating the average effect for all Mexican states, as appears to be the case here.

The findings under specification (2) are the main findings from estimating the regression model with the 9 distance decay specifications and using the traditional measurement of the magnitude of potential spatial FDI effects. Again, the findings suggest that there are no spatial FDI effects of either the intra- or inter-industry type. In all the different estimations, the spatially weighted variables of inter-regional FDI participation carry insignificant coefficients.

The findings under specification (3) are obtained from estimating the regression model with the various distance decay specifications and the alternative indicators of the scale of potential spatial FDI effects. With these alternative indicators, the findings indicate that there are spatial FDI effects. Using the first order contiguity assumption produces findings that identify a significant positive association between inter-regional inter-industry foreign participation and labour productivity of Mexican firms. The estimation based on the nearest neighbours assumption produces evidence of significant negative spatial intra-industry FDI externalities. The specification that relates spatial externalities to geographical distance in a continuous fashion produces mild evidence of positive spatial inter-industry FDI effects. In combination, these findings offer evidence that there are spatial FDI effects between the Mexican states and that these effects are of a positive and negative nature.

Finally, the findings under specification (4) incorporate the various elements of the regressions under specification (3). In this regression, I use different

Table 6.6   Intra- and inter-regional FDI externalities

	Spec(1)	Spec(2)			Spec(3)			Spec(4)
	—	Cont_1	k=2	d=2	cont_1	k=2	d=2	k=2 (interregionalFOR) Cont_1 (interregionalFIR)
Constant	0.44 (0.90)	-0.08 (0.42)	-0.09 (0.47)	-0.01 (0.42)	0.08 (0.36)	-0.51 (0.44)	0.42 (0.47)	-0.23 (0.45)
$(k)_m$	0.38 (0.05)a	0.38 (0.05)a	0.37 (0.05)a	0.37 (0.05)a	0.36 (0.04)a	0.39 (0.06)a	0.37 (0.05)a	0.39 (0.06)a
Dual	-0.16 (0.05)a	-0.16 (0.05)a	-0.14 (0.06)b	-0.04 (0.05)b	-0.13 (0.05)b	-0.12 (0.05)b	-0.13 (0.05)b	-0.13 (0.04)a
Density	0.05 (0.01)a	0.06 (0.02)a	0.06 (0.02)a	0.06 (0.02)a	0.07 (0.02)a	0.06 (0.02)a	0.06 (0.01)a	0.06 (0.015)a
Industrymix	0.52 (0.09)a	0.53 (0.09)a	0.51 (0.09)a	0.53 (0.10)a	0.54 (0.09)a	0.54 (0.10)a	0.51 (0.10)a	0.52 (0.10)a
Intra-regional For	-0.05 (0.02)a	-0.05 (0.02)a	-0.05 (0.02)a	-0.05 (0.02)a	-0.04 (0.02)b	-0.04 (0.02)b	-0.04 (0.015)a	-0.04 (0.015)a
Intra-regionalFir	0.06 (0.03)b	0.05 (0.029)c	0.06 (0.03)b	0.05 (0.027)c	0.05 (0.029)c	0.07 (0.03)b	0.05 (0.024)b	0.05 (0.024)a
Inter-regional FOR	0.05 (0.08)	0.03 (0.03)	-0.02 (0.04)	0.005 (0.03)	-0.03 (0.03)	-0.04 (0.016)a	-0.01 (0.03)	-0.04 (0.013)a
Inter-regional FIR	0.28 (0.45)	-0.02 (0.06)	-0.005 (0.04)	0.06 (0.05)	0.09 (0.04)b	-0.01 (0.02)	0.11 (0.058)c	0.05 (0.02)b
State and industry effects	Yes (0.00)	Yes (0.00)	Yes (0.00)	Yes (0.00)	Yes (0.00)	Yes (0.00)	Yes (0.00)	Yes (0.00)
Adj. R^2	0.79	0.81	0.82	0.82	0.82	0.82	0.82	0.83
N	175	175	175	175	175	175	175	175

Notes on the following page.

Notes to Table 6.6: Intra- and inter-regional FDI externalities:

(m) = Mexican-owned industry share.

Standard errors in parentheses; a, b and c indicate significance levels of 1, 5 and 10%. Estimated standard errors heteroscedastiticy-robust using Hubert/White/Sandwhich method. State and industry effects statistic is significance level of F-test whether state and industry dummies have coefficients equal to 0. Specification (1) assumes no relation between distance and spatial FDI effects. Cont_1 = first order contiguity assumption; k=2 two geographically closest neighbours; d = 2 interregional effects weighted by inverse of squared distance between state capital cities.

Specifications (1) and (2) define inter-regional foreign participation as FDI's share in regional industry and regional total manufacturing employment; specifications (3) and (4) use alternative indicators correcting inter-regional foreign participation shares for their size with respect to total FDI employment in Mexico.

I estimated the regression model for specifications (2) and (3) experimenting with all 9 different specifications of the distance matrix. In addition to the estimated effects presented in the table, none of the other spatial weights produced estimated significant effects of the inter-regional FDI variables.

Distance matrices are row standardised.

specifications for the distance matrices that capture the relation between geographical distance and inter-regional effects from intra- and inter-industry FDI participation. As mentioned earlier, it is likely that, given the variety of channels that may generate and transmit FDI externalities, the relation between distance and intra- or inter-industry FDI effects will differ. Therefore, I estimate a regression model where spatial intra-industry FDI effects are weighted based on the nearest neighbours assumption, whereas spatial inter-industry FDI effects are weighted using the first order contiguity assumption. The empirical findings presented in Table 6.6 show the variety of externality effects that are generated by foreign-owned firms. Looking first at intra-regional effects, the presence of FDI firms leads to the occurrence of negative intra-industry and positive inter-industry externalities. Furthermore, these externality effects have a multi-regional character, as the findings also contain evidence of negative spatial intra-industry and positive spatial inter-industry FDI externalities. The fact that the findings of these spatial FDI effects are based on the assumption that spatial externality processes are subject to a geographical cut-off point serves as a reminder of the importance of geographical proximity for these FDI externalities. Having said so, findings for the selected states in the previous section contain evidence of national level positive intra-industry FDI externalities, a feature which is not identified when estimating average spatial FDI effects for all 32 Mexican states.

Determinants of Spatial Externalities

To conclude the empirical analysis, I try to identify determinants of spatial FDI effects. An important debate in the literature on agglomeration economies evolves around the question whether, next to or instead of static agglomeration economies that are related to the scale or level of density of an agglomeration, there may also be dynamic place-based externality effects that are related to the composition of an agglomeration (Rosenthal and Strange 2004, Quigley 1998, Hanson 2001a). In particular, this debates centres on the question whether regional specialisation or diversity may generate positive dynamic agglomeration economies (e.g. Glaeser et al. 1992, Henderson 1997, 2003, also Combes 2000). An important implication of this discussion is that, if the composition of regional economic activity may generate agglomeration economies, it may also influence the occurrence of spatial FDI spillovers. To determine whether regional characteristics may be important for spatial FDI effects, I adapt the regression model in the following manner:

$$(6.D) \quad Y/L_{(m)} = \beta 0 + \beta_x X + \beta 6 \text{ FOR} + \beta 7 \text{ FIR} + \beta 8 \sum_{r'1;r'\neq r}^{r} W \text{ FORinterregional} +$$

$$\beta 9 \sum_{r'1;r'\neq r}^{r} W \text{ FIRinterregional} + \beta 10 \sum_{r'1;r'\neq r}^{r} W Z \text{ FORinterregional} +$$

$$+ \beta 11 \sum_{r'1;r'\neq r}^{r} W Z \text{ FIRinterregional} + \varepsilon;$$

where $X = ((k)_{(m)}, (\delta-1)L_{(m)},$ Dual, Industrymix, Density). Z contains three regional characteristics that I hypothesise to influence spatial FDI effects. First, linking the analysis to the discussion on compositional productivity effects from agglomeration, I incorporate the effects of regional specialisation and diversity into the regression model. I calculate regional specialisation as follows:

$$\text{Specialisation}_{ir} = \frac{\dfrac{(employees)ir}{(employees)r}}{\dfrac{\sum\limits_{r=1}^{32} (employees)i}{\sum\limits_{r=1}^{32} (employees)r}}$$

To calculate the level of regional diversity, I use a regional diversity indicator proposed by Combes (2000):

$$\text{Diversity}_{ir}: \frac{\left(\sum_{i'=1,i'\neq i}^{i} (employees)i'r / ((employees)r - (employees)ir)^{2}\right)^{-1}}{\left(\sum_{i'=1,i'\neq i}^{i} (employees)i' / ((employees)country - (employees)icountry)^{2}\right)^{-1}}$$

Second, I link the estimation to the discussion whether the size of the technology gap promotes or hinders FDI externalities. To do so, I include the variable GAP in vector Z of the regression model, measured as the industry level ratio of value added per employee of FDI firms over value added per employee among Mexican firms.

Table 6.7   Regional characteristics and inter-regional FDI externalities

	(1)	(2)
Intra-regional FOR	-0.04 (0.019)b	-0.06 (0.02)a
Intra-regional FIR	0.06 (0.025)c	0.05 (0.03)b
Inter-regional FOR	-0.04 (0.013)a	-0.04 (0.01)a
Inter-regional FIR	0.05 (0.02)b	0.14 (0.04)a
Inter-regional FOR*Technology Gap	–	0.02 (0.006)a
Inter-regional FIR * diversity	–	-0.07 (0.03)a
Inter-regional FOR* specialisation	–	0.02 (0.007)b
State and industry effects	Yes (0.00)	Yes (0.00)
Adj. R²	0.80	0.85
N	175	175

Notes: Standard errors in parentheses; a, b and c indicate significance levels of 1, 5 and 10%. Estimated standard errors heteroscedastiticy-robust using Hubert/White/Sandwhich method. State and industry effects statistic is significance level of F-test whether state and industry dummies have coefficients equal to 0.

Regression (1) is replicated from Table 6.6, specification (4).

Spatial weights are defined as in specification 4 in Table 6.6.

All estimations also include a constant and the variables $k_{(m)}$, Dual, Density, $(\delta-1)L_{(m)}$ and Industrymix. Initial regressions showed that interregionalFIR*GAP, interregionalFOR*Diversity and interregionalFIR*Specialization carry insignificant coefficients.

Distance matrices are row standardised.

A summary of the findings from estimating regression model (6.D) is presented in Table 6.7. The importance of the regional characteristics as determinants of spatial FDI externalities is clearly confirmed, as the interaction variables between specialisation, diversity and technology gap and intra- or inter-industry spatial FDI effects carry significant coefficients. The interaction variable between inter-regional intra-industry FDI externalities and technological differences between FDI and Mexican firms is positively associated with labour productivity of Mexican firms. Again, this finding indicates that large technological differences between FDI and Mexican firms foster the materialisation of positive FDI effects, in this case at the inter-regional level. Next, regional diversity appears to stimulate negative spatial inter-industry FDI effects. It is not entirely clear what this effect captures. It may be the case that a large level of diversity is associated with the regional presence of a large pool of suppliers, who produce inputs for FDI firms that are located in neighbouring regions. If this promotes fierce competition between these suppliers, the net result of such a situation may be the generation of additional negative FDI externalities. In contrast to the negative effect of regional diversity, regional specialisation fosters the occurrence of positive spatial intra-industry FDI externalities. This suggests that regions with specialised manufacturing activities possess a high level of absorptive capacity, facilitating the incorporation of positive productivity effects from FDI firms located in nearby regions. Finally, an additional important feature of the findings in Table 6.7 is that the significance level of the estimated effect of the original four FDI participation variables increases, indicating that the incorporation of effects from regional characteristics on spatial FDI externalities improves the overall performance of the regression model.

5. Summary and Conclusions

Contemporary research on FDI effects is developing new research strategies that incorporate the identification of the industry and geographical dimensions of FDI externalities, aiming to obtain improved indications of the full range of these externalities and to develop a better understanding of when and how these effects materialise. The industry dimension entails that FDI effects can arise both within and between industries. As for the geographical dimensions, one interpretation concerns the positive effect that agglomeration or geographical proximity can have on FDI spillovers. Another interpretation entails the recognition that there may be separate FDI effects at the national and the regional level. Finally, there is the recent acknowledgement that FDI externalities may occur within and between regions. In this chapter, I use two datasets with unpublished industry data to obtain further quantitative evidence of FDI externalities in Mexican manufacturing industries, whereby the empirical analysis contains a special focus on these industry and geographical dimensions.

The main empirical findings are as follows. With the dataset containing all 32 Mexican states, I estimate intra-regional FDI externalities, distinguishing between FDI effects within and between industries. In line with findings from previous chapters, the findings indicate that intra-industry foreign participation generates negative externalities within regions. At the same time, intra-regional inter-industry foreign participation generates positive externality effects. This part of the analysis also provides additional evidence that a large technology gap between FDI and Mexican firms fosters the occurrence of positive externality effects.

In extension of these findings, I use the dataset with detailed industry observations for selected Mexican states to estimate FDI externalities within industries and among Mexican client and supplier firms. I distinguish between national and regional level FDI participation for all these productivity effects. The findings for intra-industry foreign participation suggest the presence of positive externality effects at the national and negative externalities at the regional level. Not only does this confirm the idea that there may be separate FDI effects at the national and the regional level, it also shows that the nature of the effects may differ between these two levels of aggregation. In addition to this important finding, the importance of the regional dimension of FDI spillovers is further underlined by the finding that positive externalities among Mexican suppliers only materialise at the regional level. Furthermore, positive externalities among Mexican owned competitors, suppliers and client firms are all stimulated at the regional level in industries with a large technology gap, representing further corroborating evidence of both the regional dimension of FDI effects as well as of the positive effect of large technological differences on positive FDI externalities. In contrast to this, the estimations also identify a positive effect of industry agglomeration on the occurrence of negative FDI externalities among Mexican suppliers. This estimated effect suggests that agglomeration may foster fierce competition between these suppliers, forcing them to offer a high level of concessions to FDI client firms. Also, it may be the case that FDI firms are less willing to offer support to supplier firms if the agglomeration contains a large number of suppliers. In combination, such a situation may result in the occurrence of negative externalities through FDI backward linkages, as the findings suggest.

An alternative interpretation of the regional dimension of FDI effects is that the impact of FDI firms may differ between regions. The Mexican case offers a particularly good setting to address this issue, given the heterogeneous nature of the response by regional economies to the introduction of trade liberalisation in the late 1980s. The analysis of the dataset with selected Mexican states produces two important findings. First, there is indeed a substantial level of regional variation in the impact of FDI across the regions. Broadly speaking, the border states are benefiting from additional positive FDI externalities, linked to both intra-industry foreign participation and FDI backward linkages. In contrast to this, Mexico City in particular is subject to additional negative externalities. In combination, these findings suggest that it is very likely that FDI firms have contributed to the changes in participation shares of Mexico City and the border states in the

Mexican economy during the last two decades. Second, the findings contain further evidence that both the technology gap and industry agglomeration influence FDI effects. Large technological differences promote positive FDI externalities in most of the border states, whereas industry agglomeration fosters negative externalities in Mexico City and Jalisco. A likely explanation for the additional negative FDI externalities in agglomerated industries is that suppliers in Mexico City in particular are engaged in fierce competition, whereas at the same time FDI firms may be less willing to offer support to local supplier and client firms. The opening up of the Mexican economy has made it easier for FDI firms to source inputs from and sell products on the international market, a situation which will only have enhanced the level of competitive pressure on Mexican suppliers in Mexico City. The net result of this situation appears to be the creation of additional negative externalities in this agglomeration.

Finally, I use the dataset with the 32 Mexican states to identify spatial FDI externalities and address the question whether regional characteristics influence these effects. From estimating regression models that are based a variety of assumptions on how geographical distance may influence the occurrence of spatial FDI effects, I find evidence of negative intra-industry and positive inter-industry spatial FDI externalities. Importantly, these effects only materialise when I use the improved indicators of the scale of potential inter-regional FDI effects. The specifications of the distance decay effect suggest that spatial FDI externalities are subject to a geographical cut-off point, indicating the importance of geographical proximity in externality transmitting processes. In addition to this, the findings also indicate that regional diversity fosters negative spatial inter-industry FDI externalities, whereas both regional specialisation and large technological differences between FDI and Mexican firms enhance positive spatial FDI effects of an intra-industry nature.

Chapter 7

Foreign Direct Investment and Backward Linkages with Local Suppliers: Survey Evidence from Nuevo León

1. Introduction

Quantitative research on FDI effects in host economies has produced an impressive amount of evidence on the existence and nature of FDI externalities. Having said so, it is also the case that the evidence is characterised by a substantial degree of heterogeneity, indicating the complexity of identifying these productivity effects. Furthermore, it is important to recognise that quantitative studies face limitations when it comes to answering important questions on why and how FDI externalities are transmitted. For instance, both Haskel et al. (2007) and Javorcik et al. (2004) present evidence indicating that the nationality of FDI firms may be important for the materialisation of positive externalities. Although such a finding is clearly important, it does not tell us why this may be the case. Why would say US-owned FDI firms generate less or more externality effects compared to EU FDI? Especially from the point of view of a host economy's government, it is vital to understand what forces are captured by such a broad characteristic of nationality, if it wants to design policies to influence FDI productivity effects.

The closest to identifying actual mechanisms of FDI externalities have come empirical studies that estimate the effect of industry foreign participation on productivity of domestic firms in supplying industries, as these effects can be assumed to be related to the externality-transmitting channel of backward linkages. Having said so, in addition to the fact that the evidence on the nature of the externalities that this channel may transmit is heterogeneous, findings of a positive association between industry foreign participation and productivity of firms in supplying industries represent only indirect evidence of the workings of FDI backward linkages. Research of a more qualitative nature, usually consisting of detailed case studies or small scale surveys, are more informative on this issue, providing detailed accounts of the extent and nature of local sourcing of FDI firms in host economies. The downside of this approach is that the evidence of spillovers tends to be circumstantial (Blomström and Kokko 1998), the evidence is usually based on small non-representative samples of FDI firms and domestic producer firms and domestic suppliers tend to be omitted from the analysis altogether.

The purpose of the present chapter is to conduct an empirical study on the extent and nature of FDI backward linkages with local suppliers, to complement the quantitative analysis on FDI externalities of the previous chapters. I conduct the analysis of FDI local linkages in the state Nuevo León, representing one of the main agglomerations of economic activity within Mexico. For the empirical analysis of inter-firm linkages between FDI and Mexican firms, I rely on unique data obtained via firm level surveys that I designed and carried out in Nuevo León in 2000-2001. One dataset is based on survey responses from a representative sample of both foreign-owned and Mexican producer firms. The second dataset is based on a survey among a representative sample of the entire pool of Mexican suppliers in the region. To the best of my knowledge, the analysis presented in this chapter represents the first study that is representative for both foreign-owned and domestic producer firms as well as for domestic suppliers.

The chapter is constructed as follows. In section 2 I briefly review research on FDI and backward linkages in host economies. In section 3 I present some background information on the state Nuevo León and I discuss the research design. This section also presents statistical comparisons between FDI and Mexican producer firms regarding the extent and nature of their linkages with local suppliers. Also, I use information from the supplier survey to assess the importance of the support that is provided by producer firms. Section 4 presents a statistical analysis of determinants of the level of use of local suppliers and the extent to which producer firms offer support to their suppliers, estimating the effect of firm characteristics such as age, size, type of ownership and maquiladora status. Finally, section 5 summarises and concludes.

2. Foreign Direct Investment and Local Linkages

Research on FDI backward linkages in host economies focuses on several issues. One issue is that the presence of FDI firms can generate positive externalities by demanding improvements in production processes of their local suppliers (Katz 1969, Javorcik and Spatareanu 2005). For instance, Alfaro and Rodríguez-Clare (2004) interviewed several managers of domestic firms in Costa Rica and found that the pressure exercised by FDI firms had been an important factor fostering improvements in the production process of these domestic firms. Potter et al. (2002) present similar findings from a survey in the UK among FDI and their UK suppliers.

Another issue is the extent to which FDI firms use their suppliers and the nature of their buyer-supplier relations. Dunning (1993) reviews several studies and concludes that this extent varies considerably between host economies. Also, although it is difficult to make generalisations, especially given the absence of information on the extent of local sourcing by domestic firms, the perception is that FDI firms tend to be integrated to a lower degree compared to their domestic counterparts. As for the nature of backward linkages, Lall (1980) presents detailed information on a variety of types of support provided by a foreign-owned truck manufacturer in India. For instance, the FDI firm was actively involved in helping with the creation of new suppliers, provided support for the technological upgrading of the production processes of several suppliers, gave advice on improving their sourcing processes and offered financial assistance in some cases. Potter et al. (2002) present evidence from their survey in the UK of several types of support provided by FDI firms, including assistance with production, quality testing and training. Smarzynska and Spatareanu (2005) present further corroborating evidence of supportive linkages between FDI and domestic firms in the Czech Republic and Latvia (see also Javorcik 2008). Related evidence on supportive FDI firms for several developing countries is presented in UNCTAD (2001), Ivarsson and Alvstam (2005) Dunning (1993) and Hallbach (1989).

The third issue concerns attempts to relate FDI characteristics to the level of use of local suppliers. For instance, Belderbos et al. (2001) find that Japanese FDI firms that produce for host economy markets use more local inputs compared to Japanese firms producing for international markets. Dunning (1993) reports findings of a similar nature from several other studies. Another factor that may be important is the age of the FDI firm, as it takes time and familiarity with the host economy to develop successful sourcing linkages (Driffield and Noor 1999, UNCTAD 2001, Hallbach 1989, Giroud and Hafiz Mirza 2006). Also, the role of a foreign-owned plant within its MNE is likely to affect the degree to which the plant can make autonomous sourcing decisions (Zanfei 2000). Other factors that may be important include the size of a FDI firm, the type and nature of its main production technologies and the mode of establishment (UNCTAD 2001).

FDI and Backward Linkages: Evidence from Mexico

There is a considerable amount of evidence on the extent to which FDI firms use Mexican suppliers. For instance, findings from a survey among 63 foreign-owned manufacturing firms indicate that one out of three firms sourced more than 25 percent of their inputs in Mexico (UNCTC 1992). Another more recent survey of similar size among FDI from EU member countries shows a level of local content of about 35 percent (Martínez-Solano and Phelps 2003). Similar percentages are reported by Moran (2005) and from company case studies by Shaiken (1990) and Ivarsson and Alvstam (2005). Having said so, other studies, including Altenburg et al. (1998), Fuentes et al. (1993), Dussel (1999) and Padilla-Pérez (2008) all report that FDI firms use Mexican suppliers to only a very moderate degree.

Up until recently, a deciding factor in the process of local sourcing has been whether a FDI firm operates in the maquiladora programme. Several studies that have looked into sourcing behaviour of maquiladora firms conclude that the level of local integration of these firms is disappointingly low (Silvers 2000, Buitelaar and Padilla-Pérez 2000, Biles 2004, Kenney and Florida 1994, Brannon et al. 1994). Maquiladora firms are focused mainly on obtaining production costs advantages, by locating labour intensive production processes in Mexico, whilst benefiting from tax breaks on imported inputs and re-exported assembled products (Sklair 1993). The development of local linkages with Mexican suppliers has not been a goal of the maquiladora programme, and maquiladora firms appear to have shown little interest in increasing their level of use of these suppliers.

However, the current composition of firms that participate in the maquiladora programme is very different from two decades ago. In the last 20 years, there has been a marked growth of technologically more advanced maquiladora firms (Wilson 1992). A good example of this change is the production of televisions in the border states, which has evolved from pure assembly operations in the early years to more sophisticated production activities in recent years (see Carrillo and Mortimore 1998). To capture this changing nature of maquiladora firms, the terminology first, second and third generation maquiladora firms has been introduced (Carrillo and Hualde 1998, Buitelaar and Pérez 2000, Sargent and Matthews 2004). First generation maquiladora firms are characterised by labour intensive production processes and are geared solely towards maximising the benefits from relative low labour costs in Mexico. Second generation maquiladora firms incorporate a larger share of actual production instead of only assembly operations and rely more on the use of skilled labour. Third generation maquiladora firms are characterised by a high level of autonomy within their MNE and are responsible for most decisions regarding products, production processes and sourcing (Buitelaar and Padilla-Pérez 2000, Carrillo and Gomis 2003). An example of a third generation maquiladora firm is Delphi-General Motors, analysed by e.g. Carrillo and Hualde (1998) and Lara and Carrilllo (2003). As a result of the existence of several types of maquiladora firm, it may no longer be the case that all maquiladora firms will use Mexican suppliers to such a low degree as they have become known for.

In addition to studies on the extent to which FDI firms use local suppliers, some studies have also looked at the nature of FDI backward linkages. For instance, the earlier-referred-to study by UNCTC (1992) found several FDI firms to be actively involved in helping their local suppliers, whereby information and advice on how to improve production processes was the most common type of support. Altenburg et al. (1998) present detailed information on a range of types of support offered by a small number of FDI firms in the car and clothing industries. Their findings indicate that quality control measures and training programmes were common types of support in the car industry. Less frequent was assistance in the form of the provision of special tools and assistance with the sourcing of inputs. Importantly, as Dussel (1999) emphasises, while FDI firms may offer support, at the same time they expect that local suppliers also make investments to improve their performance. Additional evidence on how Ford and Volvo have improved their local supplier base is presented by Carrillo (1995) and Ivarsson and Alvstam (2005).

Looking at the evidence on the extent and nature of FDI backward linkages with Mexican suppliers, there are several limitations that constrain the empirical findings. First, in addition to the fact that the evidence on the level of use of local suppliers is very diverse, there is no evidence on the extent to which Mexican producer firms use local suppliers, making it impossible to assess whether FDI firms actually differ from domestic firms in their level of use of these suppliers.

Also, it is usually not clear what the sampling procedure has been to select FDI firms, raising questions to what degree the findings are biased or representative of the population of FDI firms at the regional or national level. Furthermore, evidence on the existence of different types of maquiladora firms, in relation to the issue of local sourcing, is purely qualitative of nature. It may be true that there are more modern maquiladora firms that source more local inputs compared to first generation maquiladoras, but there is no statistical evidence supporting this claim. In a similar fashion, evidence on supportive linkages between FDI and Mexican suppliers is based on data obtained mainly from case studies. Finally, the evidence on supportive linkages is one-sided and biased towards the opinion of foreign-owned firms, as there is no consistent analysis of the overall importance and effects of supportive linkages among Mexican suppliers that receive the support.

3. Research Design and Comparisons between FDI and Mexican Producer Firms

The aim of the empirical analysis in this chapter is three-fold. First, I want to obtain indications of the extent to which FDI firms actually use local suppliers. Second, I want to obtain information on the existence of supportive linkages between FDI and local suppliers and determine which types of support are the most common, as supportive linkages can be linked to the occurrence of positive externality effects. Third, I want to identify firm level characteristics that influence both the level of use of Mexican suppliers and the level of supportiveness of producer firms.

To address these questions, I analyse the operations of FDI firms in the state Nuevo León, located in the North East of Mexico. The reason for selecting this state is three-fold. First, as the findings in Chapter 3 indicate, FDI is attracted to agglomerations within Mexico, indicating the importance of studying the operations and effects of foreign-owned firms in such agglomerations. The state Nuevo León constitutes the second largest agglomeration of economic activity after Mexico City and has been very successful in participating in processes of economic liberalisation and trade promotion since the 1980s (Vellinga 2000, 1995, The Economist 1998, Jordaan and Harteveld 1997). Second, Nuevo León has been a favourable destination region for FDI for some time (Jordaan and Harteveld 1997). It contains a substantial level of foreign participation in its manufacturing sector, offering ample opportunities to study the operations and effects of FDI firms. Third, as the findings in Chapter 6 indicate, there is evidence of positive externalities in Nuevo León (and other border states) that are associated with the existence of FDI backward linkages, underlining the importance of obtaining more information on the extent and nature of these linkages in this agglomeration.

Table 7.1 presents information for 1998 on the composition of the manufacturing sector of Nuevo León and the state's participation in Mexico's manufacturing activities. Overall, Nuevo León employs more than 320,000 workers in its manufacturing sector, representing almost 8 percent of Mexico's total manufacturing

Table 7.1   Employment manufacturing industries in Nuevo León: 1998

Activity	Number of employees	Share in manufacturing sector Nuevo León (%)	Share in manufacturing sector Mexico (%)	Included in producer survey
Total manufacturing	321,085	n.a.	7.8	–
Food	32,621	10.1	5.1	–
Beverages and tobacco	7,356	2.3	5.0	–
Inputs for textiles industries	2,804	0.8	2.1	–
Production of textiles	2,485	0.8	4.4	–
Clothing	18,627	5.8	3.7	–
Leather, leather products and synthetic leather	3,874	1.2	2.5	–
Wood and wood products	2,596	0.8	3.3	–
Paper and paper products	9,931	3.0	12.0	–
Printing	7,462	2.3	7.8	–
Products derived from petroleum	4,755	1.5	10.1	–
Chemical industries	15,468	4.8	7.4	xxx
Plastics and rubber	15,631	4.9	8.0	xxx
Non metallic minerals	30,400	9.5	15.2	–
Basic metals	12,528	3.9	16.7	–
Metal products	42,852	13.3	14.8	xxx
Machinery and equipment	18,623	5.8	19.2	xxx
Computers, telecommunication equipment and related electronic products	14,920	4.6	5.7	xxx
Electric generators and related products	23,408	7.3	13.2	xxx
Cars, engines and related products	30,486	9.5	7.1	xxx
Furniture	14,354	4.5	8.5	–
Other	9,904	3.0	7.7	–

Source: Based on data taken from Economic Census 1999.

work force. Looking at the relative importance of the various activities, the industries of machinery and equipment, basic metals, electric generators and related electrical products, metal products, non metallic minerals and paper and paper products all employ more than 12 percent of the national work force of these industries. To select industries from which to select producer firms for the survey, I considered their share in Mexico's aggregate manufacturing sector as well as the importance of the industries for the economy of Nuevo León. As indicated in the last column of Table 7.1, this led to the selection of chemical industries (chemical industries and plastics and rubber), machinery and equipment (machinery and equipment, metal products), electric and electronic industries (computers etc. and electric generators) and the car industry (cars, engines and related products).

For these industries, I generated a list of both Mexican and foreign-owned producer firms that have at least 150 employees. Using information from a variety of sources, including the local branch of INEGI, the local affiliate of the American Chamber of Commerce, the state government, local industry associations and business directories, I compiled a list of 180 manufacturing firms. Based on this extensive search, as well as additional information obtained from interviewing several key actors in the regional economy, I feel confident that the list contains the majority of FDI and Mexican producer firms in the selected industries. All the firms are located in the metropolitan area of Monterrey, which is the capital city of the state and constitutes the main agglomeration of economic activity in this region. In the summer of 2000, I carried out a pilot study, interviewing 30 FDI and Mexican producer firms. Following this pilot study, the remaining 150 firms were contacted and visited by interviewers to participate in the survey which I developed following the pilot study. Another 52 firms agreed to participate in the survey, resulting in a total of 82 participating firms, which represents a response rate of 46 percent.

Table 7.2 presents a range of firm level characteristics of both Mexican and foreign-owned producer firms. FDI firms are significantly larger and younger. Also, they employ a different mixture of personnel, as the shares of technicians and engineers in their workforce are significantly larger than in Mexican firms. There are also differences in forward linkages: Mexican firms sell a significantly larger share of their production on the domestic market, whereas FDI firms are more engaged in exporting activities.

Next, Table 7.2 presents several indicators of the sourcing patterns of FDI and Mexican firms. The average local content of material inputs of the two types of firm does not differ significantly from the sample average of 26 percent. There are also no significant differences between FDI and Mexican firms regarding the use of suppliers in the US, the extent to which they use sophisticated inputs and the extent to which they are involved in the production of inputs in-house.[1] One aspect where the firms do differ is the level of sourcing from Mexican suppliers that are located in Mexican states other than Nuevo León, as Mexican firms report a significantly higher use of these suppliers. Finally, there is a significant difference in the extent to which both types of firm purchase production services from local providers. This type of input is often not included in surveys on local sourcing by FDI firms. Production services refer to situations where a producer firm sends an

---

1   Sophisticated inputs are defined in the survey as inputs that are not directly available 'off the shelf' and require specialised technologies to produce.

Table 7.2   Main characteristics of foreign-owned and Mexican producer firms

Characteristics	Mean	Mexican	Foreign-owned	Significance (p-value)
Age of plant	25.05	30.65	21.46	0.07 (c)
Number of employees	590	306.31	771	0.02 (b)
% engineers	7.89	5.15	9.64	0.03 (b)
% technicians	16.25	8.23	21.38	0.001 (a)
% production workers	62.48	74.59	54.73	0.001 (a)
Material inputs as % of production costs	61.16	60.87	61.37	0.91
% of sales in Nuevo León	19.93	23.97	17.35	0.15
% of sales in Mexico	34.75	47.30	26.72	0.002 (a)
% of sales in US	5.25	24.48	48.42	0.003 (a)
% of material inputs produced in-house	15.82	15.59	15.80	0.97
% of material inputs sourced in Nuevo León	26.05	32.22	22.10	0.13
% of material inputs sourced in rest of Mexico	18.35	28.08	12.13	0.005 (a)
% of material inputs sourced in US	35.12	29.53	38.69	0.16
% of material inputs that are sophisticated	18.24	15.59	19.88	0.54
Local production services as % of production costs	8.864	12.03	6.61	0.02 (b)

Note: a, b and c indicate significance levels of 1, 5 and 10%; t-test for equality of means.

Source: Survey among local producer firms in Nuevo León, 2000.

intermediate input to a local provider of a production service. After this provider has treated the intermediate input, it is sent back to the producer firm to be incorporated again into the production process. Treatments include metal plating and metal stamping, plastic moulding, surface conversion and coating, assembly activities and finishing activities. According to the survey responses, Mexican firms make significantly more use of these local production services than FDI firms do. The importance of including this type of input in the survey is supported by the finding that the sample average share of local production services in total production costs is about 9 percent. Also, only seven out of the 82 firms indicate that this type of service is not relevant for their production process.[2]

---

2   The importance of including this type of input became apparent during the pilot study among producer firms, as several firms indicated that although this type of input is usually not included in surveys on local sourcing, it may in fact represent a non-negligible share of production costs. The findings from the producer survey confirm this.

In addition to the survey among producer firms, I also carried out another survey among Mexican supplier firms in the region. Here I faced an important problem as to how to find out which firms were operating as actual supplier to the producer firms. Information provided by the producer firms to select suppliers would generate a biased sample. To avoid this, I compiled a second list of firms, containing all firms in the region that employ less than 150 employees and are classified as operating in the manufacturing sector. To compile this list, I collected contact information and addresses for all manufacturing firms that were registered with local industry associations, resulting in a list of 1,100 firms. I treated this large set of firms as the pool of potential suppliers to producer firms.

During the summer of 2000, I organised a telephone survey among this pool of potential suppliers to identify firms that were operating as supplier to the producer firms in the selected industries. After a three month period, all firms had been contacted. Of the large list of firms, 356 agreed to participate in the telephone survey, representing a response rate of about 33 percent. Table 7.3 presents the main responses to the telephone survey from the 300 firms that indicated that they sell to local producer firms. One important finding is that 84 percent of the suppliers indicate that sales to the producer firms are very important or important for their sales, indicating the importance of local sourcing by the producer firms for the local economy. A majority of the suppliers supply mainly to Mexican firms, another 28 percent indicate that FDI and Mexican producer firms are equally

Table 7.3   Main findings telephone survey among suppliers

How important are local producer firms for your sales?	Very important	Important	Somewhat important	Not important
	45.7%	39.4%	10.9%	4%
Who are your main clients among local producer firms?	Mexican firms	FDI firms	Both equally important	n.a.
	62%	10%	28%	
What type of product do you supply?	Parts and components	Raw materials	Machinery (replacement) parts	Production services
	20.9%	14.6%	21.7%	18%
Type of firm	Manufacturing firm	Distributor	Workshop	n.a.
	60%	9%	31%	

Note: N = 300.

Source: Telephone survey among suppliers in Nuevo León, 2000.

important clients and 10 percent of the suppliers have FDI firms as main client. It is also interesting to note that there is a considerable variety in the types of input that are produced, including parts and components, raw materials, (replacement) machinery parts and production services. As for the types of supplier, the majority of firms classify themselves as manufacturing firm, another 30 percent consist of workshops and the remaining 9 percent act mainly as distributor.

Support for Local Suppliers

Following the telephone survey, I applied a second firm level survey among the group of firms that are operating as actual supplier to the producer firms. In the autumn of 2000, I carried out a pilot study by interviewing 20 local suppliers. During the spring of 2001, the remaining 280 firms were approached by interviewers to participate in the supplier survey. In the end, 100 out of the group of 300 firms participated in either the pilot study or the survey, representing a response rate of 33 percent.

The survey among local suppliers provides important information on the prevalence and importance of support that is provided by producer firms. The main responses are presented in Table 7.4. The majority of suppliers indicate that their production processes have improved or largely improved as a result of support provided by the producer firms, suggesting that backward linkages between the producer and supplier firms have generated positive externality effects. Furthermore, there is a significant difference between the group of firms that supply to Mexican producer firms and the group of firms with FDI clients: the share of firms in the second group that indicates that being a supplier to local producer firms has largely improved their production processes is significantly larger, suggesting that being a supplier to a FDI firm leads to larger positive externalities than being a supplier to a Mexican firm does.

Next, there is no significant difference between suppliers of Mexican and FDI firms in terms of the sources of improvement of their production processes, as the producer firms are an equally important source of improvement for both groups of supplier. Interestingly, and in contrast to findings for other host economies by e.g. Smarzynska and Spatareanu (2005) and Alfaro and Rodríguez-Clare (2004), pressure exercised by producer firms towards their suppliers to improve their production processes is a much less important source of improvement. Instead, the suppliers indicate that independent investments have been far more important. Additional information obtained from the pilot study among local suppliers indicates that suppliers perceive such independent investments to be an integral part of their functioning as supplier to the producer firms. Having said this, as the investments are made in response to the presence of the producer firms, the resulting improvements are a form of positive externalities.

The last two questions concern the importance of the support for the suppliers and the extent to which local suppliers can use support provided by one client firm to improve their functioning for other client firms. For both questions, there is a

Table 7.4　Overall effects on suppliers

Main Clients of Local Suppliers	To what extent has your production process improved by being supplier to producer firms?					Sign.
	Largely	*Improved*	*Somewhat*	*The same*	*Decreased*	
Mexican and/or FDI as main client	43.9	45.6	7	1.8	1.8	(b)
Mexican client	9.3	67.4	20.9	2.3	0	
Main source of improvement?						
	Support of local clients	*Pressures from local client*	*Independent investments*			
Mexican and/or FDI as main client	38.6	12.3	49.1			
Mexican client	46.5	11.6	41.9			
How important is the technological and operational assistance for your firm to be an efficient supplier?						
	Very important	*Important*	*Somewhat important*	*Not important*		
Mexican and/or FDI as main client	56.1	38.6	5.3	0		(b)
Mexican client	37.2	53.5	9.3	0		
Can the technological assistance you receive from one client be used to improve your production for other clients?						
	Very much	*A little*	*No*			
Mexican and/or FDI as main client	89.5	8.8	1.8			(a)
Mexican client	60.5	37.2	2.3			

Note: N = 100; a and b indicate significance levels of 1 and 5%; Kendall's tau-b statistic.

Source: Survey among local suppliers, Nuevo León, 2000-2001.

significant difference between the suppliers of the two groups of producer firm. The share of firms among suppliers with FDI clients that indicates that the support has been very important to become an efficient supplier is significantly larger than the share of firms among suppliers of Mexican producer firms. Also, the share of suppliers that indicates that the support offered by one client firm can be used to improve their functioning for other client firms is significantly larger in the group of suppliers with FDI client firms. This finding is particularly important, as it constitutes a strong indication that positive externalities are likely to be linked to this support. Recalling the discussion on externalities through backward linkages in Chapter 2, positive externalities only materialise when support offered by a producer firm outweighs concessions that a supplier is expected to make in return. The fact that 90 percent of suppliers of FDI firms indicate that they can use support provided by one producer firm to improve their supply relations with other firms strongly suggests that the effects of the support outweigh concessions that they need to make in return.

The producer survey provides more detailed information on the types and frequency of support that producer firms provide their local suppliers with. In the survey, I distinguished between ten types of support that fall broadly into two categories, depending on the extent to which they are directly linked to improving the production process of local suppliers. Types of support that have a relative direct impact are the supply of blue prints, lending of machinery, provision of special tools, direct assistance with technical production and quality control and the provision of training programmes. Support with a less direct link to the suppliers' production processes include assistance with the creation of new suppliers, assistance with general business and organisation development, financial assistance, assistance with sourcing and cooperation in the development of new inputs. The responses of the producer survey are presented in Table 7.5.

Looking first at the findings for support offered to suppliers of material inputs, FDI firms are most frequently involved in helping suppliers with production processes and quality control, with 75 percent of the foreign-owned producer firms indicating that they provide this type of support frequently. Second in importance is the provision of blue prints, followed by assistance in the creation of new suppliers, the supply of special tools and training programmes and assistance in the development of new inputs. General business support, financial assistance and the lending of machinery occur least frequently; having said this, more than 30 percent of the FDI firms is engaged in these latter types of support frequently.

Comparing the frequencies of support between FDI and Mexican producer firms, there are strong indications that FDI firms are more supportive. In particular, the share of FDI firms that indicates to be involved frequently in the provision of blue prints, machinery, special tools, training of personnel of suppliers and assistance with production and quality control is significantly larger than the share among Mexican producer firms. To make sure that this difference in findings is not caused by the presence of young FDI firms that may be more supportive to their local suppliers in their early years of operation, I also compare frequencies

Table 7.5   Types of support offered by FDI and Mexican producer firms

Type of assistance	Support offered to suppliers material inputs			Support offered to suppliers production services		
	FDI	*Mature FDI*	*Mexican*	*FDI*	*Mature FDI*	*Mexican*
Blue prints	71.8 (a)	75 (a)	14.8	65.7 (a)	71.4 (a)	11.5
Machinery	25.6 (a)	30 (b)	3.7	17.1 (b)	19 (c)	7.7
Special tools	59 (a)	65 (a)	11.1	45.7 (a)	47.6 (a)	11.5
Technical production and quality control	84.6 (c)	75	66.7	91.4 (a)	85.7 (b)	53.8
Training personnel	59 (a)	65 (a)	22.2	45.7 (b)	42.9 (b)	19.2
Create new suppliers	61.5	65	59.3	45.7	42.9	34.6
General business and organisation development	35.9	40	25.9	31.4	33.3	19.2
Grants/loans/accelerated payment	33.3 (b)	30 (b)	11.1	22.9	14.3	19.2
Assistance in sourcing	48.7	40	33.3	n.a.	n.a.	n.a.
Cooperation in developing new inputs	56.4	70	63.6	n.a.	n.a.	n.a.

Notes: N = 82; a, b and c indicate significant levels at 1, 5 and 10%; Kendall's tau-b statistic.

Table shows percentage of firms that indicate that they offer a particular type of support frequently.

Mature FDI firms are firms that have been in operation in Nuevo León for at least 15 years.

Source: survey among producer firms in Nuevo León, 2000.

of support between mature FDI and Mexican producer firms. As the responses in Table 7.5 indicate, there is no bias from the presence of young FDI firms, as mature FDI firms are also significantly more supportive than Mexican producer firms for the same types of support, except for the provision of quality assistance.

The second set of columns in Table 7.5 presents similar information on the various types of support that are offered to local providers of production services. Comparing the frequencies of the various types of support between the two types of producer firm, FDI firms are again significantly more supportive, for largely the same types of support as is the case with support offered to suppliers of material inputs. Furthermore, the finding that FDI firms are more supportive is not caused by the presence of a group of generally more supportive young FDI firms, as the same significant differences in supportiveness also apply to mature FDI firms versus Mexican producer firms.

4. Determinants of the Level of Use of Local Suppliers and the Provision of Support

The dichotomous comparisons between FDI and Mexican firms indicate that there is no significant difference in the level of use of suppliers of material inputs, Mexican firms use significantly more local production services and FDI firms are significantly more supportive. These dichotomous comparisons do not control for the effect of other firm level characteristics, however. To obtain further statistical evidence on which factors are important determinants of the level of backward linkages, I specify the following two regression models:

(7.A) $LC_i = \beta 0 + \beta_x X_i + Industry_i + \varepsilon_i$;

(7.B) $LPS_i = \beta 0 + \beta_x X_i + Industry_i + \varepsilon_i$

LC_i stands for the level of use of local suppliers of material inputs by firm i. LPS_i represents the level of use of local providers of production services by firm i. Xi is a vector of firm characteristics that I hypothesise to influence the level of use of both types of supplier. The variable Industryi controls for structural differences in local sourcing patterns across the different industries and εi is the error term of the estimation. The variables are listed and defined in Table 7.6. LC is measured as material inputs purchased from suppliers in Nuevo León as share of total material input costs. LPS is measured as locally purchased production services as share of total production costs.

The first control variable that I include is the age of the producer firm, measured as the number of years that a firm has been in operation in Nuevo León. Based on findings from previous studies, I expect a positive effect of this variable on the level of use of local suppliers, as it takes time to develop successful and sustainable linkages with local suppliers (UNCTAD 2001, Dunning 1993, Driffield and Noor 1999). In addition to this age variable, I also include a variable labelled Mature, to control for the possibility that mature firms have a structurally different sourcing behaviour. Mature firms have been in the region since the era of import substitution, which may have resulted in these firms having developed a different use of local suppliers compared to firms that have started operations more recently during the period of trade liberalisation.

The next variable is size, measured as the total number of employees in 2000. I expect to find a negative effect of this variable on the use of local suppliers, as large firms have more resources and skills to produce inputs themselves in-house (Dunning 1993). Also, large firms are likely to have a scale of demand for inputs that exceeds the capacity of the average-sized supplier. This means that the pool of potential suppliers for large firms in general is smaller, which lowers the likelihood that suppliers with a sufficient production capacity will be found locally.

I include two variables that are related to ownership type and type of producer firm. The variable FDI is a dummy variable that captures whether a producer firm

Table 7.6   Definition of variables

Name	Description	Measurement
LC	Use of local suppliers	Material inputs purchased locally as share of total material inputs
LPS	Use of local production services	Local production services as share of total production costs
Age	Age of firm	(ln) number of years in operation
Mature	Mature firms	1 if age >=15, 0 otherwise
Size	Size of firm	(ln) number of employees in 2000
FDI	Ownership of firm	1 if >=10% of total assets is owned by foreign company, 0 otherwise
Maqui	Maquiladora firm	1 if firm is registered in maquiladora programme, 0 otherwise
USinputs	Use of US suppliers	Inputs sourced from US as % of total material input costs
USsales	Sales to US	Sales to US as % of total sales
Parts and Components	Parts and components	Parts and components as % of total material input costs
Assembly	Assembly style production	1 if production consists only or mainly of assembly activities, 0 otherwise
Industry	Industry dummies	Car industry, electric and electronics industry, chemical industries, machinery and equipment

Note: All variables are calculated with data from the producer survey.

is foreign-owned. Following common practice, I classify a firm as foreign-owned when more than 10 percent of its assets are in foreign hands. Based on the findings in the previous section, I expect that this variable has no effect on the level of use of local suppliers of material inputs and a negative effect on the local purchasing of production services. Second, I include the variable Maqui, which is a dummy variable capturing whether or not a producer firm is operating in the maquiladora programme. It is not clear what to expect from this variable. On the one hand, the traditional impression of maquiladora firms is that they are integrated poorly into their local economy. On the other hand, maquiladora firms of the second or third generation may not differ in their use of local suppliers from other firms.

The variables USinputs and USsales control for the effect of the level of international backward and forward linkages with supplier and client firms in the US. I measure the level of international backward linkages as the percentage of total material input costs that is sourced from the US. The level of international forward linkages is calculated as the percentage of sales that is made on the US market. I expect a negative effect on the level of use of local suppliers from both these variables, given existing evidence for other host economies (Dunning 1993, Belderbos et al. 2001, UNCTAD 2001).

The variables Parts and Components and Assembly control for the effect of the style of production on local sourcing. The effect of Parts and Components, measured as the share of ready made parts and components in total material input costs, may be positive or negative. On the one hand, it may be that firms that use many ready made inputs have an incentive to use local suppliers intensively, to benefit from proximity advantages. On the other hand, it may be that such firms have a policy to rely on established suppliers located elsewhere, to avoid problems with delivery, quality and price that may occur when using local Mexican suppliers. The variable Assembly is a dummy variable indicating whether or not the production process of a producer firm consists only or mainly of assembly activities. I expect that this variable has a negative effect on the use of local suppliers.

Finally, I include industry dummies to capture all the effects that are related to structural differences in local sourcing between industries. I also estimate the regression models allowing for clustered standard errors at the municipality level to control for municipality effects.

The main findings from estimating regression model (7.A) are presented in Table 7.7. As the dependent variable is restricted to the interval [0,1], I use double censored Tobit regression techniques (see Wooldridge 2002). The first column with results contains the findings from estimating the model with the control variables age, size, maquiladora status and type of ownership. According to these findings, it appears that the only important factor is whether or not a firm is mature, which has a negative effect on the use of local suppliers.

When I add interaction variables between maturity and foreign ownership or maquiladora status to the estimation, the results are substantially different. The interaction variables control for the possibility that mature FDI and maquiladora firms, which have been in operation since the period of import substitution, are characterised by a different level of local integration compared to firms that started operations during trade liberalisation. As the findings in column two indicate, the maquiladora status of a firm has two opposing effects. Maquiladora firms use significantly more local suppliers of material inputs, except for mature maquiladora firms which use significantly less of these suppliers. These findings suggest that mature maquiladora firms consist of first generation maquiladora firms, which are well known for their low level of local integration. The estimated positive effect of maquiladora status suggests that this effect captures the presence of second and third generation maquiladora firms, which are notably well integrated into the local economy. Importantly, the estimated effect of type of ownership remains insignificant, indicating that there are no differences in the level of use of local suppliers between Mexican and foreign-owned producer firms.

The third column adds the additional control variables that capture the effects of international backward and forward linkages and the type of production process. There is a significant negative effect from the level of use of suppliers located in the US, which is as expected. Furthermore, the level of international forward linkages also lowers the use of local suppliers. A likely explanation for this estimated effect is that product specifications for inputs of products that are sold on the

Table 7.7   Determinants of the use of local suppliers of material inputs

	1	2	3	4
	LC	LC	LC	MC
Age	-0.07 (0.07)	-0.07 (0.06)	-0.02 (0.06)	-0.008 (0.06)
Mature	-0.25 (0.13)b	-0.07 (0.17)	-0.16 (0.16)	-0.15 (0.16)
Size	-0.022 (0.03)	-0.013 (0.03)	-0.05 (0.035)	-0.02 (0.03)
FDI	0.02 (0.07)	-0.10 (0.22)	0.26 (0.21)	0.24 (0.20)
Mature * FDI	–	-0.11 (0.15)	0.16 (0.15)	0.19 (0.15)
Maqui	0.014 (0.07)	0.34 (0.16)b	0.45 (0.22)b	0.40 (0.22)b
Mature * Maqui	–	-0.25 (0.15)c	-0.31 (0.13)b	-0.19 (0.14)
FDI * Maqui	–	0.05 (0.15)	0.005 (0.015)	-0.26 (0.14)c
Usinputs	–	–	-0.16 (0.03)a	-0.20 (0.003)a
Ussales	–	–	-0.06 (0.03)b	-0.09 (0.03)a
Parts and Components	–	–	0.02 (0.009)b	-0.04 (0.08)
Assembly	–	–	-0.14 (0.07)b	0.05 (0.07)
Industry fixed effects	Yes (0.00)	Yes (0.00)	Yes (0.00)	Yes (0.00)
AIC	74.751	77.739	41.339	32.558
LL	-29.375	-27.896	-15.780	-10.405
LR	8.395	11.406	37.265	47.085
Pseudo R^2	0.09	0.13	0.41	0.55
N	82	82	82	82

Notes: Estimated standard errors in parentheses. Standard errors robust to heteroscedasticity and clustering at the municipality level. a, b and c indicate significance level of 1, 5 and 10%.

Industry fixed effects statistic is for F test whether industry dummies have coefficients equal to 0.

LC is level of use of suppliers in Nuevo León; MC is the level of use of suppliers located in Mexico.

international market are more stringent. This may discriminate against the use of local Mexican suppliers, when international suppliers outperform them in terms of quality, cost competitiveness and reliable supply systems.[3] Producer firms that use a large volume of ready made parts and components use significantly more local suppliers, suggesting that these firms are trying to benefit from proximity effects by using these suppliers. Additional information from the producer survey shows that firms that use a relative large amount of ready made parts and components use mostly standardised inputs that are specific to their industry, suggesting that local suppliers are particularly important in providing this type of input. In contrast to this, the extent to which the production process is characterised by assembly style operations is negatively associated with the level of use of local suppliers.[4] Finally, the last column presents the findings from replacing the dependent variable with the level of use of suppliers located in Mexico (i.e. suppliers in Nuevo León and suppliers located in other Mexican states). Again, maquiladora firms source a significantly larger percentage of their total material inputs from Mexican suppliers. There is no additional negative effect among mature maquiladora firms; in this case, foreign-owned maquiladora firms are reporting a significantly lower use of Mexican suppliers. In addition to this, the findings also indicate that the levels of international backward and forward linkages are both negatively associated with the level of use of suppliers in Mexico.

The findings from estimating regression model (7.B) with double censored Tobit regression techniques are presented in Table 7.8. Again, the initial findings from estimating the regression model with age, size, type of ownership and maquiladora status are disappointing. The only variable with a significant coefficient is ownership type, and, according to findings in the previous section, the sign of the estimated effect is incorrect. Adding the interaction variables between maturity and foreign ownership or maquiladora status improves the results considerably, as shown in the second column with findings. The results indicate that mature firms use significantly more local providers of production services. Having said so, mature FDI firms source significantly less from these providers. Again, maquiladora status carries a significant positive coefficient. There is no evidence of a negative effect among mature maquiladora firms; this time it is foreign-owned maquiladora firms that are using significantly less local providers.

Column three contains the findings from adding the additional control variables. The positive effect of maquiladora status and maturity are confirmed,

---

3   Additional information from the producer survey supports this explanation, as producer firms indicate that issues of cost competitiveness and reliability are common problems that they experience with local Mexican suppliers.

4   I also experimented with a variable that captures the level of autonomy of a producer firm regarding sourcing decisions (e.g. Brannon et al. 1994, Zanfei 2000). For the full sample of firms, this variable carries an insignificant coefficient. Estimating the regression model for only FDI firms produces a significant positive effect of the level of autonomy in sourcing decisions on the level of use of local suppliers.

Table 7.8   Determinants of the use of local providers of production services

	1	2	3	4
Age	0.02 (0.03)	0.001 (0.02)	0.02 (0.02)	0.09 (0.03)a
Mature	0.04 (0.05)	0.11 (0.05)b	0.10 (0.06)c	0.17 (0.09)b
Size	-0.009 (0.01)	0.007 (0.01)	-0.004 (0.01)	0.01 (0.01)
FDI	0.06 (0.025)b	-0.09 (0.067)	0.02 (0.07)	0.07 (0.09)
Mature * FDI	–	-0.09 (0.045)b	0.05 (0.06)	0.10 (0.85)
Maqui	0.02 (0.02)	0.18 (0.07)b	0.21 (0.08)b	0.43 (0.09)a
Mature * Maqui	–	-0.07 (0.047)	-0.04 (0.04)	-0.12 (0.07)c
FDI * Maqui	–	-0.12 (0.048)b	-0.22 (0.05)a	-0.27 (0.07)a
Usinputs	–	–	0.01 (0.04)	0.02 (0.02)
Ussales	–	–	0.01 (0.01)	-0.02 (0.02)
Parts and components	–	–	-0.002 (0.003)	-0.02 (0.003)a
Assembly	–	–	-0.04 (0.026)c	-0.06 (0.03)b
In-house inputs	–	–	–	0.04 (0.01)a
Industry fixed effects	Yes (0.00)	Yes (0.000	Yes (0.00)	Yes (0.00)
AIC	-113.711	-126.076	-90.391	-54.631
LL	63.856	73.038	60.195	42.316
LR	9.836	20.546	23.012	38.774
Pseudo R^2	0.12	0.24	0.34	0.70
N	82	82	82	50

Notes: Estimated standard errors in parentheses. Standard errors robust to heteroscedasticity and clustering at the municipality level. a, b and c indicate significance level of 1, 5 and 10%.

Industry fixed effects statistic is F test whether industry dummies have coefficients equal to 0.

as is the negative effect among foreign-owned maquiladora firms. The estimated effect of mature FDI becomes insignificant. There is no evidence that the other variables are important, suggesting that there are structural differences between decisions to use local suppliers of material inputs and local providers of production services. The exception to this is the estimated mildly significant negative effect of assembly style production. The explanation for this effect is most likely that firms that specialise in assembly activities have little need for production services that provide some form of treatment of intermediate products.

In light of the insignificance of several of the control variables, I re-estimate the regression model adding an additional control variable that captures the extent to which a producer firm is involved in the production of inputs in-house.[5] The drawback of adding this variable is that several producer firms did not provide a usable response to the question on this issue in the survey on this issue, resulting in a decrease in the number of observations. Keeping this in mind, the last column shows the findings from estimating the regression model with the additional control variable. The inclusion of the additional variable has several effects. The estimated effect of age becomes significant and positive, indicating that older firms use significantly more local production services. Maturity also maintains its positive effect. Importantly, the findings now confirm the impression that mature maquiladora firms use significantly less local suppliers, which is evidence of the presence of first generation maquiladora firms. Maquiladora status itself is positively associated with the level of use of local providers, indicating that second and third generation maquiladora firms are very well integrated into the local economy. Foreign-owned maquiladora firms source significantly less from these providers. Also, the variables Parts and Components and Assembly now both carry a significant negative effect, reflecting the relative low need for production services among firms with a high reliance on ready made inputs and assembly style production. The estimated effect of the level of in-house production of inputs is positively associated with the use of local production services, which reflects the need for such services among producer firms that are involved themselves in the production and treatment of inputs. Finally, the estimated effect of ownership type remains insignificant. This is in contrast to findings from the dichotomous comparison between FDI and Mexican firms in the previous section, indicating that, when controlling for the effect of other firm level characteristics, ownership type does not affect the level of use of local providers of production services.

Determinants of Support

Similar to the analysis of determinants of the level of use of local suppliers of material inputs and local providers of production services, I can estimate a regression model to obtain statistical evidence on which producer firm characteristics influence the level of support that is provided to supplier firms. To do so, I specify the following regression model:

5　I measure this variable as material inputs produced in-house as percentage of total material input costs.

(7.C)   Supporti = $\beta 0 + \beta z\ Zi$ + Industryi $+\epsilon i$;

where Supporti is a binary variable taking the value of 1 if firm i offers support frequently to their local suppliers and 0 otherwise, Zi is a vector of firm characteristics, Industryi contains the industry effects and ϵi is the error term of the estimation.

To be able to estimate a similar regression model for the various types of support that are offered to both suppliers of material inputs and providers of production services, I include the following variables in Zi: age, size, FDI, FDImature, Maqui, Maquimature, USinputs, LC and LPS. The variables are as defined in Table 7.6. As for the types of support, I focus on the provision of blue prints, the supply of machinery, the provision of special tools and the training of personnel of suppliers. These types of support are identified in the producer survey as occurring most frequently; also, dichotomous comparisons between FDI and Mexican producer firms indicate that FDI firms are more supportive for these types of support. Given the binary nature of the dependent variable, I estimate regression model (7.C) with logit regression techniques.

The main empirical findings are presented in Table 7.9. The first set of columns contains the empirical findings for determinants of support that is offered to local suppliers of material inputs. One variable that carries a significant positive coefficient for all four types of support is firm size, which can be explained by the fact that large firms are more likely to possess sufficient resources to offer these types of support on a frequent basis. In addition, the age of a producer firm has a positive effect on the provision of blue prints and training, suggesting that familiarity with the region enhances supportiveness.

The estimated effect of foreign ownership is positive for all four types of support. This finding is important, as it indicates that FDI firms are significantly more supportive, even when controlling for other firm level characteristics. In addition to this, the interaction variable between maturity and foreign ownership is also positively associated with the provision of support, indicating that, among FDI firms, mature firms are more supportive.

The three maquiladora related variables show a mixture of effects. Maquiladora status has a significant positive effect on all four types of support, indicating that maquiladora firms are significantly more supportive. At the same time, mature maquiladora firms are significantly less supportive, which again suggest that this effect concerns first generation maquiladora firms. The interaction variable between foreign ownership and maquiladora status also carries a positive coefficient, indicating that among maquiladora firms foreign-owned firms are more supportive.

The effect of the level of international backward linkages is mixed. On the one hand, firms with a large dependence on US suppliers of material inputs provide less support in the form of blue prints. On the other hand, they provide significantly more support in the form of the provision of special tools. The negative effect suggests that producer firms that are less dependent on local suppliers are less

Table 7.9   Determinants of support

Variables	Support offered to suppliers of material inputs				Support offered to providers of production services			
	Blue prints	Machinery	Special tools	Training	Blue prints	Machinery	Special tools	Training
Age	4.79 (2.15)a	0.09 (1.11)	1.45 (1.54)	1.14 (0.57)b	0.97 (0.50)b	1.77 (1.78)	0.29 (0.85)	0.52 (.31)c
Size	1.86 (1.06)c	0.63 (0.26)c	0.45 (0.17)a	0.70 (0.40)c	0.13 (0.36)	-0.36 (0.46)	0.27 (0.26)	0.65 (0.39)c
FDI	27.47 (12.28)a	35.03 (3.37)a	41.04 (5.07)a	1.63 (0.58)a	16.88 (2.28)a	14.84 (6.07)a	1.64 (1.98)	0.02 (1.81)
FDI mature	11.41 (5.93)a	17.78 (1.66)a	-0.36 (2.08)	2.44 (1.29)b	2.14 (1.44)	3.16 (2.52)	-0.16 (1.23)	1.20 (1.03)
Maqui	62.36 (6.96)a	71.50 (2.54)a	18.94 (4.44)a	5.25 (2.32)a	24.41 (2.56)a	21.24 (0.68)a	17.68 (2.42)a	4.57 (1.25)a
Maqui mature	-23.56 (3.10)a	-18.76 (1.93)a	2.67 (2.14)	-2.36 (1.35)c	-18.84 (0.69)	-18.72 (3.14)a	-0.10 (1.42)	-2.64 (0.76)a
FDI*Maqui	16.91 (6.72)a	33.92 (0.67)a	19.83 (2.14)a	0.70 (0.68)	0.09 (0.10)	15.56 (3.13)a	19.41 (2.40)a	-0.12 (1.01)
Inputs from US	-6.68 (2.69)a	1.39 (1.11)	7.87 (1.66)a	0.48 (1.42)	0.90 (1.49)	-0.50 (2.12)	2.21 (1.67)	0.96 (1.51)
LC	2.66 (2.93)a	4.45 (1.92)b	7.02 (1.74)a	-0.86 (1.31)	—	—	—	—

Table 7.9 continued   **Determinants of support**

Variables	Support offered to suppliers of material inputs				Support offered to providers of production services			
	Blue prints	Machinery	Special tools	Training	Blue prints	Machinery	Special tools	Training
LPS	—	—	—	—	-4.84 (2.74)c	12.76 (3.99)a	1.91 (3.56)	-3.91 (4.57)
Industry fixed effects	Yes (0.00)	Yes (0.00)	Yes (0.00)	Yes (0.00)	Yes (0.00)	Yes (0.00)	Yes (0.00)	Yes (0.00)
AIC	60.344	55.322	59.031	83.881	47.1502	35.226	67.995	77.015
LL	-15.172	-13.661	-16.515	-28.940	-18.75	-12.61	-25.997	-29.508
Pseudo R2	0.64	0.53	0.59	0.31	0.50	0.44	0.32	0.26
N	82	82	82	82	82	82	82	82

Notes: Estimated standard errors robust to heteroscedasticity and clustering at the municipality level.

Dependent variable is whether a firm offers support to its suppliers on a frequent basis.

Industry fixed effects statistic is F test whether industry dummies have coefficients equal to 0.

a, b and c indicate significance levels of 1, 5 and 10%.

willing to offer support to these suppliers. The positive effect may indicate that producer firms with a relative large reliance on US suppliers may be engaged in supportive processes among local suppliers in an attempt to lower their dependence on non-local suppliers.

Finally, the estimated effect of the level of use of local suppliers on the level of supportiveness is significant and positive for three of the four types of support. Of course, there is a problem of endogeneity here, as firms that offer more support may be using more local suppliers, and vice versa. Accepting this caveat, this finding is important as it indicates that the level of use of local suppliers and the level of supportiveness appear to have a positive effect on each other.

The second set of columns of Table 7.9 presents the findings on determinants of the provision of support to local providers of production services. In contrast to the findings for support to suppliers of material inputs, the size and age of a producer firm do not affect the level of support towards these local providers. Type of ownership is important when it comes to providing support in the form of blue prints and the lending of machinery; again, foreign-owned producer firms are significantly more supportive. There is no additional positive effect among mature FDI firms, however.

Again, the effect of the maquiladora variables is mixed. Maquiladora firms are significantly more supportive for all four types of support. The estimated effect of the interaction variable between foreign ownership and maquiladora status indicates again that foreign-owned maquiladora firms are more supportive, when it comes to providing support in the form of the lending of machinery and special tools. Mature maquiladora firms are significantly less supportive, reflecting the low level of local integration of first generation maquiladora firms.

The estimated effect of international backward linkages is insignificant. This is in line with the findings on determinants of the level of use of local providers of production services, confirming that international backward linkages are not important when looking at the decision making process surrounding the purchase of local production services. Interestingly, the level of use of local providers of production services does not have a uniform effect on the level of supportiveness towards these providers. For instance, when it comes to the provision of blue prints, the findings indicate that producer firms are less supportive when they use a relative large amount of local production services. In contrast, the level of use of local providers has a positive effect on the provision of machinery, whereas there is no effect when it comes to the provision of special tools or training. Compared to the clear positive relation between the level of local sourcing of material inputs and the level of support that is provided to the suppliers of these inputs, the relation between the level of use and the level of supportiveness is much less straightforward when it concerns local providers of production services.

5. Summary and Conclusions

Quantitative research on FDI externalities faces limitations when it comes to answering questions such as why and how these FDI effects are transmitted. The closest have come studies that focus on productivity effects from FDI firms to domestic firms in supplying industries, as such inter-industry effects can be linked to the channel of backward linkages. Having said so, the evidence on the importance of this channel remains indirect, and empirical evidence on the nature of externalities that appear to be transmitted through this channel is heterogeneous. In light of this, coupled with the fact that findings in previous chapters suggest that FDI backward linkages form an important externality-transmitting channel in Mexican manufacturing industries, the purpose of this chapter is to present a detailed analysis of the level and nature of linkages between FDI firms and local suppliers in Nuevo León, which represents the second largest agglomeration of economic activity in Mexico. For this analysis, I have taken great care to avoid pitfalls and shortcoming of previous research on FDI linkages, by applying surveys to representative and unbiased samples of both FDI and Mexican producer firms and Mexican suppliers of inputs.

The main findings can be summarised as follows. The sample average of the level of use of local suppliers is 26 percent of total material inputs. There is no significant difference in the level of use of suppliers of material inputs between FDI and Mexican producer firms. Most studies on FDI linkages only look at the level of use of suppliers by FDI firms, assuming that this level of use is very likely to be lower compared to domestic firms. The survey findings indicate that this assumption is clearly incorrect for the producer firms in Nuevo León, which underlines the importance of analysing the level of local integration of both types of producer firm to assess whether FDI firms do actually differ from domestic firms. As for the use of local providers of production services, a type of input that is often not included in surveys on local sourcing, the dichotomous comparison between FDI and Mexican firms suggests that the latter type of firm uses more of these services. This difference disappears when I control for the effect of other firm level characteristics, indicating that type of ownership has no effect on the level of use of local suppliers of material inputs and production services.

Second, the surveys provide important evidence on the prevalence and types of support that are provided by the producer firms. Overall, a majority of suppliers indicate that an important part of their sales is made to local producer firms, indicating the importance of local demand for inputs by producer firms for the local economy. Also, most suppliers indicate that their production processes have benefited substantially from support provided by producer firms. Furthermore, findings from the supplier survey indicate that FDI firms are significantly more supportive, suggesting that FDI firms generate more positive externality effects than Mexican producer firms do. This impression is further supported by the finding that suppliers with FDI clients are better able to use support received from one client firm to improve their functioning for other clients. Turning to the producer survey, the findings confirm that FDI firms are significantly more supportive to both suppliers of material inputs and production services, especially when it comes to providing support that has a direct link with improving the production processes of these suppliers. Furthermore, the overall impression is that producer firms are supportive in general, as the findings show that types of support that occur the least frequent are still provided by more than 30 percent of the producer firms.

Third, the statistical analysis of determinants of the use of local suppliers of material inputs and production services produces several interesting findings. As mentioned earlier, type of ownership does not affect the level of use of these suppliers, indicating that FDI and Mexican firms are locally integrated to the same degree. In contrast to this, maquiladora status has a significant influence on the level of use of local suppliers. Producer firms that are registered as operating in the maquiladora programme source significantly more from local suppliers, except for mature maquiladora firms which source significantly less local inputs. In combination, these findings suggest that there are both first generation and second and third generation maquiladora firms in the manufacturing sector of Nuevo León with distinct different levels of local content. First generation maquiladora firms conform to the traditional image of maquiladora firms that use local suppliers to only a limited extent, whereas second and third generation maquiladora firms are noticeably well integrated into the regional economy. This constitutes important statistical evidence for previous case study findings that suggest that modern maquiladora firms have a larger positive impact on their local economy. In addition to this central finding, levels of international forward and backward linkages have a negative effect on the use of local suppliers of material inputs, whereas a high reliance on ready made parts and components leads to a higher use of local suppliers of mainly standardised inputs. Assembly style production lowers the use of both suppliers of material inputs and production services, whereas the level of in-house production of inputs by producer firms enhances the use of local production services.

Finally, I conduct statistical analysis to identify firm level characteristics that influence the level of supportiveness of producer firms, focusing on determinants of those types of support that are directly linked to improving the production processes of local suppliers. The findings indicate that type of ownership has a significant effect on the level of supportiveness of a producer firm: controlling for other firm level characteristics, FDI firms are significantly more supportive. Furthermore, the importance of second and third generation maquiladora firms for the regional economy is underlined by the finding that these maquiladora firms are also significantly more supportive, suggesting that both FDI firms and second and third generation maquiladora firms are a good source of positive externalities. Again, first order maquiladora firms are significantly less supportive, which is in line with the traditional image of maquiladora firms. Other firm level characteristics that appear to be important for some types of support include the size and the age of a producer firm, which both have a positive effect. Interestingly, whereas the level of use of suppliers of material inputs has a clear positive relation with the level of support that the producer firms provide these suppliers with, this relation is not apparent when looking at the relation between the level of use of providers of production services and the level of supportiveness of producer firms to these suppliers.

FDI Externalities in Mexican Manufacturing Industries: Main Findings, Policy Recommendations and Future Research

1. Introduction

The economic impact of FDI is a widely researched topic. In line with the growing interest in the role of externalities in processes of economic growth, the last two decades have seen a rapidly growing body of quantitative research that tries to identify and quantify FDI externalities in a wide variety of host economies. Despite the large amount of evidence on these FDI effects, this research field is facing several important challenges. One challenge concerns the continuing debate on the question how important FDI externalities are, and whether these effects are of a positive or negative nature. Cross-country growth regressions provide little evidence in support of a general positive FDI growth effect, whereas the body of evidence from micro-economics research on FDI externalities in individual host economies is very heterogeneous in nature. Although recent empirical studies have broadened their scope, both by increasing the number of host economies for which FDI spillovers are estimated, as well as by estimating for FDI effects within and between industries, the large degree of heterogeneity of the evidence has not diminished. Another important challenge concerns the question whether there are factors that affect the materialisation and perhaps also the nature of FDI externalities. The most commonly accepted factor is the level of absorptive capacity of domestic firms, which points at the requirement that these firms must possess sufficient knowledge and skills to be able to absorb new technologies from FDI firms. The empirical translation of this concept of absorptive capacity into the technology gap between FDI and domestic firms, which is interpreted as a direct inverse indicator of the level of absorptive capacity of domestic firms, is problematic as it rests on an incomplete interpretation of the underlying catch up thesis. The findings from several studies that show that a large technology gap may be conducive rather than detrimental to positive FDI externalities further indicate the need to address this problem.

In a continued pursuit to develop improved identification strategies of FDI externalities, a number of recent studies have started to explore spatial dimensions of these spillover effects. Although these studies contain only few direct links to the literature on spatial dimensions of externality effects, the literature on the geography of externalities contains strong indications that such spatial dimensions of FDI effects may be important. Geographical concentration of economic activity generates agglomeration economies that are uniquely linked to the existence of the agglomeration. Given the large extent of similarity between the channels that underlie FDI externalities and the mechanisms of agglomeration economies, the hypothesis that agglomeration will influence FDI externalities is easily proposed. Also, several studies identify a clear regional dimension of knowledge spillovers in the context of estimating regionalised versions of the knowledge production function, suggesting that estimations of FDI effects may benefit from distinguishing between national level and regional FDI externalities. Finally, related research on the spatiality of knowledge spillovers suggests that a full capturing of FDI externalities in a host economy requires the estimation of both intra- and inter-regional FDI effects.

The empirical evidence on spatial dimensions of FDI externalities is promising, but also reflects that there is an important scope for improvement. For instance, the clear hypothesis that agglomeration may affect FDI externalities has been addressed by only a surprisingly small number of studies. Findings on the regional dimension of FDI externalities are diverse, indicating that such a dimension is far from automatic. Having said so, it appears that several of the studies that estimate for such a regional dimension fail to control for all types of regional foreign participation. This problem also applies to studies that estimate FDI spillovers between regions. Most of the studies on inter-regional effects from FDI can also be criticised for relying on ad-hoc specifications of the possible relation between geographical distance and spatial FDI externalities, as well as for using imprecise indicators of the magnitude of spatial intra- and inter-industry FDI effects. In combination, these issues underline the need to apply consistent and careful estimations to ensure that the full range of effects from foreign participation in a host economy is captured.

The purpose of this book is to develop a variety of empirical estimations of the spatial dimensions of FDI externalities, in the context of estimating FDI externalities in Mexican manufacturing industries. The need to obtain new empirical evidence on the operations and effects of FDI firms in this host economy is indicated by the fact that existing evidence on FDI effects in Mexico is based on the analysis of data for the 1970s and 1980s, a period characterised by strict policies of import substitution and government intervention. The drastic change in development strategy towards economic liberalisation and trade promotion in the late 1980s led to a dramatic increase in inward FDI into the Mexican economy, making foreign-owned activity vital to the success of this new development strategy. In this context, new evidence on FDI effects for more recent periods is required to improve our understanding of FDI activities in the contemporary Mexican economy. Furthermore, the period of trade liberalisation is characterised by far-reaching locational changes of economic activity within Mexico, offering a very interesting setting to analyse the spatial dimensions of FDI externalities.

2. Discussion of Main Empirical Findings

Overall, the empirical findings contain strong indications that the presence and operations of FDI firms generate significant externality effects in the Mexican manufacturing sector. For instance, the cross-sectional estimation of intra-industry FDI externalities produces evidence of positive FDI effects, with the aid of the new instrument that solves for the problem that the variable measuring intra-industry foreign participation is endogenous to the estimated regression model. Contrary to the common notion that OLS estimations are biased upwards as FDI firms gravitate towards high productivity industries, the findings for Mexico indicate that FDI firms gravitate towards industries with low levels of labour productivity. This particular type of bias is caused by the fact that FDI firms concentrate in labour intensive industries. Controlling for this tendency, the IV estimations identify substantial positive intra-industry FDI externalities at the national level.

Having said so, the findings also indicate that FDI firms are generating both positive and negative externality effects. For instance, whereas I identify positive intra-industry FDI externalities at the national level, at the regional level this type of industry foreign participation generates negative externality effects among Mexican firms. Also, the estimations of intra-industry FDI externalities with the regional dataset for all 32 Mexican states, as well as the empirical study on determinants of regional growth, produce evidence that FDI firms generate negative externality and growth effects at the intra-regional level. One explanation for this finding may be that the negative competition effect that FDI firms can generate has a regional dimension. Alternatively, the estimated negative effect may reflect a situation where the presence of FDI firms drives up prices of regional inputs, resulting in a decrease in efficiency among Mexican firms.

In extension of these findings on intra-industry FDI externalities, the estimations also provide important indications on inter-industry FDI effects. By and large, the findings on inter-industry FDI externalities suggest that these effects are of a positive nature, whereby the estimations with the sample of industries for selected states indicate that FDI backward linkages are the main channel of these effects. The case study on input-output linkages between producer firms and suppliers in the state Nuevo León provides important additional information on this. Its findings indicate that support provided by producer firms is important for generating improvements in production processes of these suppliers, suggesting the materialisation of positive externality effects. The analysis of the responses of the producer survey indicates not only that most producer firms are supportive to a substantial extent, it also shows that foreign-owned producer firms are significantly more supportive than Mexican producer firms. As for the level of use of local suppliers, there is no difference between Mexican and FDI producer firms. The majority of studies on FDI linkages reflect the assumption that FDI firms are likely to be less integrated into their local economy compared to domestic firms. The empirical findings in Chapter 7 indicate that this is not the case for producer firms in Nuevo León, as FDI and Mexican firms show similar levels of use of local suppliers. Therefore, externality effects from FDI in Nuevo León are linked to the feature that foreign-owned firms are significantly more supportive compared to Mexican producer firms.

As indicated earlier, there are important problems with research that uses the technology gap as direct inverse indicator of the level of absorptive capacity of domestic firms. This interpretation of the technology gap is insufficiently linked to the underlying catch up thesis on the importance of international flows of technology between leading and lagging countries as stimulator of economic growth in the latter group of countries. The catch up thesis points at the importance of two elements: a sufficient level of absorptive capacity in the lagging countries, and a sufficient potential scope of positive externalities, captured by the level of technological differences between the two groups of countries. This indicates that the interpretation of the technology gap as direct inverse indicator of the level of absorptive capacity is flawed, as it muddles together these two main elements of the need for sufficiently large technological differences and a sufficient level of absorptive capacity.

The empirical findings in this book represent a strong rejection of the interpretation of the technology gap as direct inverse indicator of the level of absorptive capacity of domestic firms. I estimate the effect of the technology gap between Mexican and FDI firms on FDI externalities with various regression models, using national level industries, industries in selected states and with the regional dataset for all 32 Mexican states. In all these estimations, I find that large technological differences between FDI and Mexican firms stimulate rather than hinder positive FDI externalities. A good example of this important finding can be found in the empirical analysis of Chapter 6, which finds that, at the regional level, positive intra-industry FDI externalities, as well as externalities through FDI backward linkages and FDI forward linkages, are all stimulated in industries with a large technology gap. Clearly, these findings challenge the notion that the technology gap can be interpreted as reflecting the inverse level of absorptive capacity of domestic firms. Instead of this interpretation, I believe that a positive relation between the technology gap and FDI externalities reflects the importance of a sufficiently large scope of potential externalities. A large scope of potential spillovers forms an important incentive to Mexican firms to engage in externality-facilitating investments, thereby improving their level of absorptive capacity, which allows them to benefit from the new knowledge and technologies incorporated in FDI firms. Also, it is likely that negative competition effects are absent or less severe when the level of technological differences between domestic and FDI firms is sufficiently large. This last point is supported by the finding that although FDI firms generate negative intra-industry externalities at the regional level, regional industries with a large technology gap enjoy additional positive intra-industry FDI effects. This suggests that in these industries, where the negative competition effect is absent or less severe, there is scope for positive externalities to materialise.

Turning to the spatial dimensions of FDI externalities, the empirical chapters provide a variety of empirical evidence in support of the notion that these dimensions can play an important role. As discussed earlier, the large degree of similarity between the channels of FDI spillovers and mechanisms of agglomeration economies strongly suggests that agglomeration or geographical proximity between firms can influence FDI spillovers. The findings from the quantitative estimation of determinants of FDI location behaviour in Mexican regions during the 1990s underline the importance of addressing this hypothesis. I find ample evidence that, controlling for the effects of a number of other regional characteristics, FDI firms are attracted to regions within Mexico that contain agglomerations of manufacturing and input providing activity. Given the preference of FDI firms for operating in such regions, the relation between agglomeration and FDI externalities is clearly important to investigate.

The estimations of FDI externalities with the sample of national level manufacturing industries produce findings that indicate that agglomeration enhances positive FDI externalities, as positive FDI effects tend to occur in industries with a relative high degree of agglomeration. Findings for the selected states confirm this positive effect for both intra- and inter-industry externalities. The qualification to this is that this effect appears to apply in particular to industries that are characterised by a simultaneous high level of agglomeration and a large technology gap. Having said so, the estimations for the selected states also find that agglomeration may promote negative externality effects between FDI firms and Mexican suppliers. In extension of this, in relation to the heterogeneity of the FDI impact across states, I find that, depending on the region, agglomeration can foster positive or negative FDI effects. For the border states, I find that agglomeration enhances positive FDI effects of both an intra- and inter-industry nature. In contrast, the findings for Mexico City in particular indicate that this region is subject to additional negative FDI externalities in industries with a relative high level of agglomeration. An explanation for such a negative effect may be that agglomeration leads to an upward pressure of prices of regional inputs. Also, the agglomeration of Mexico City is likely to have a large pool of suppliers, which can lead to fierce competition between these firms. This may force these firms to offer more concessions to FDI firms to secure demand for their products. The opening up of the Mexican economy will have only increased the level of competitive pressure on these suppliers, as it has become easier for FDI firms to import inputs from the US and other countries. Not only may this force Mexican suppliers to offer more concessions, it may also lower the willingness of FDI firms to offer support to these suppliers, given the large and growing pool of Mexican and international suppliers. In combination, this may have led to the creation of negative externalities among Mexican suppliers, which is captured by the estimated positive effect of industry agglomeration on negative FDI externalities. An important implication of the existence of such regional heterogeneity in the externality impact of FDI is that it is very likely that FDI firms have contributed to the structural locational changes that have occurred in the Mexican economy following the introduction of trade liberalisation, whereby the border states have experienced substantial increases in their participation in the Mexican economy, in strong contrast to the decreasing share of Mexico City. Another implication of the findings is that the concept of agglomeration should be interpreted as a multi-faceted phenomenon, incorporating several processes that may stimulate positive or negative FDI externalities, depending on the regional context.

Finally, the estimations produce important evidence on the existence and nature of spatial FDI growth and externality effects. The empirical study on determinants of regional growth in Chapter 4 identifies both positive and negative spatial FDI growth effects. The negative spatial growth effect may represent the occurrence of negative FDI externalities that transcend state borders. Alternatively, it may indicate that regions with a relative large amount of regional FDI attract investment flows from regions nearby, thereby lowering the growth rate of these regions. The findings in Chapter 6 show that the estimated effect of intra-regional foreign participation is similar for the analysis using detailed industry observations of the selected states and for the analysis of the regional dataset for all 32 Mexican states. In both cases, the estimations identify negative intra-industry and positive inter-industry intra-regional externalities. Furthermore, the estimations with the regional dataset produce findings that indicate the presence of significant inter-regional FDI effects. Similar to intra-regional effects, spatial intra-industry FDI effects are negative, whereas spatial inter-industry FDI externalities are of a positive nature. The estimations of spatial FDI effects also indicate that it is very important to pay special attention to the relation between geographical space and externalities and to the measurement of the magnitude of potential spatial FDI effects. Given the variety of channels that may underlie FDI externalities, it seems unrealistic to assume that all these effects are affected by geographical distance in an equal manner. Therefore, to increase the likelihood that spatial FDI effects are identified, estimations should experiment with several specifications of this relation between distance and spatial externalities. The estimations of Chapter 6 identify spatial FDI effects only when I use various specifications of the possible relation between FDI externalities and geographical space and alternative indicators of the scope of spatial FDI effects. The findings that confirm the presence of spatial intra- and inter-industry FDI externalities also indicate that there is a negative effect of distance on these spatial externalities, confirming the importance of geographical proximity between firms for the materialisation of these effects.

3. Policy Implications

The empirical findings that I present in this book have several important implications for policymaking at both the state and federal level in Mexico. The first policy implication relates to the effect of the technology gap on FDI externalities. Following the interpretation that the technology gap is a direct inverse indicator of the level of absorptive capacity of domestic firms, government policies should focus on attracting new FDI firms that are not too dissimilar from Mexican firms. However, the findings presented in this book indicate that positive externality effects are stimulated when the technology gap is large. This means that, instead of attracting FDI firms that are not too dissimilar from Mexican firms, policies should aim to attract FDI firms that are technologically substantially more advanced than Mexican firms. This will increase the potential scope of positive externality effects and also lower the chance that negative competition effects arise. A deciding factor for the success of this policy will be whether domestic firms can increase their absorptive capacity to absorb new technologies and knowledge incorporated in FDI firms. Therefore, policies to attract technologically advanced FDI firms should be accompanied by policies that help Mexican firms to engage in externality-facilitating investments, thereby enhancing the likelihood that Mexican firms will benefit from positive externality effects.

Interestingly, this policy implication can also be related to the presence of maquiladora firms of different generations in the Mexican economy. The well-known criticism of this type of FDI firm is that, besides the generation of large numbers of manufacturing jobs, these firms have a limited impact on the Mexican economy. However, the findings in Chapter 6 show that the border states, states that contain a large share of maquiladora activity, benefit from additional positive externalities in industries with large technological differences. This can be taken as evidence of positive externality effects that modern second and third generation maquiladora firms generate. Further corroborating evidence comes from the producer survey in Nuevo León. As the survey findings indicate, second and third generation maquiladora firms are significantly more integrated into the local economy and also offer a large amount of support to their suppliers, which can be linked to the materialisation of positive spillover effects.

Another policy implication follows from the findings on FDI location behaviour. The findings show that the regional presence of agglomerations of Mexican and foreign-owned manufacturing firms and suppliers of material inputs enhance the probability that a region is selected by new FDI firms. The importance of this is further underlined by the findings that suggest that both FDI activity and agglomeration economies have been important drivers of regional growth during the last 20 years. This suggests that policies aimed at developing or maintaining the regional presence of an agglomeration of economic activity will not only promote the attractiveness of the region for new FDI firms but also stimulate the generation of externality-related growth effects. This hints at the presence of a process of cumulative causation, where policies to attract new FDI firms can lead to further economic growth of the agglomeration, which in turn serves to attract more foreign owned activity, etc.

Having said this, the findings on the effects of FDI firms, in combination with the structural changes of the Mexican economy following the introduction of trade liberalisation, suggest that the issue may be more complex though. The process of agglomeration and FDI location must be seen within the context of wider spatial adjustments of the Mexican economy, where economic activity has come to be concentrated in a limited number of production centres in the border states and in Mexico City. In this context, policies that focus on ensuring that the preconditions of a regional economy are conducive for economic growth may be more successful. For instance, the costs of producing in the border states are increasing, due to the high level of agglomeration of economic activity. This may generate opportunities for other states to attract manufacturing firms from the border states in the future. An important aspect that will need to be addressed is whether and to what extent such policies will require multi-regional or federal funding. For instance, the findings of negative spatial growth effects from regional FDI suggest that an increase in FDI in a given region may hurt other regions by drawing away investment flows from these regions. Also, there is the danger that, when several regional governments are competing for new FDI firms, they end up in a position where they are forced to offer large concessions that offset the positive impact of new FDI activity. In relation to this, it is important to recognise that FDI firms that are willing to relocate to regions that do not have a substantial agglomeration of economic activity are likely to consist of types of firm that have limited business relations with their local environment, lowering the potential positive impact that such FDI firms can have on their local economy. In light of these issues, the need for multi-regional and federal policy coordination and funding seems clear.

A third policy implication relates to the findings that indicate that FDI spillover effects may be of a positive or negative nature. Not only does this indicate that there is no guarantee that policies designed to attract new FDI firms will automatically lead to the occurrence of positive FDI externalities, it also points at the possibility that an increased presence of FDI firms may have a negative effect on Mexican firms. This applies to both intra- and inter-industry FDI effects. As discussed earlier, the fact that the impact of FDI is subject to regional heterogeneity indicates that FDI externalities can be influenced by their regional context. This means that policies that aim to attract new FDI firms not only need to consider the possibility that FDI firms can generate negative instead of positive externality effects, the success of such policies also depends on a thorough analysis of the likely impact of the regional context on these FDI effects. The best example of the importance of this is that the empirical findings indicate that agglomeration enhances positive or negative FDI effects, depending on whether we look at FDI effects in the border states or in Mexico City. Clearly, the relation between agglomeration and FDI effects is complex and region specific, requiring careful analysis.

Finally, the feature that FDI externalities have spatial dimensions will need to be recognised explicitly for successful policy making. For instance, the empirical study on drivers of regional growth during the last 20 years clearly indicates that positive and negative growth effects from both agglomeration and FDI activity are not confined at the intra-regional level. The estimations of inter-regional intra- and inter-industry FDI externalities in Chapter 6 corroborate this. The presence of inter-regional FDI effects clearly indicates the need for policy coordination at the multi-regional or federal level. In a situation where federal-level policy making in Mexico has remained underdeveloped and modest in scale at best, the existence of these multi-natured inter-regional FDI effects calls for policy funding, implementation and coordination that transcends state borders. Assessing the variety of empirical findings in this book, successful government policies will need to promote and facilitate investments that improve preconditions for regional growth (e.g. investments in regional infrastructure or schooling), design policies that attract the right type of FDI with respect to enhancing the potential scope of positive externalities within specific regional contexts, provide support to enhance the level of absorptive capacity of Mexican firms and implement measures that support positive and curb detrimental regional and spatial externality effects from agglomeration and FDI. Such a combination of policies will only be successful when the right combination of regional, multi-regional and federal funding and coordination is adopted.

4. Future Research on FDI Externalities

The empirical findings presented in this book contain several indications concerning future research questions and directions, both for research on FDI in Mexico and for the general research field on FDI externalities. Focusing first on Mexico, it is clear that more empirical research on FDI effects is called for. First, given the large increase in FDI firms following the creation of NAFTA in 1994, it will be particularly interesting and important to see how intra- and inter-industry FDI externalities have developed under growing levels of economic liberalisation and trade promotion in recent years.

Second, more research is needed to obtain more evidence and a better understanding of the spatiality of the externality impact of foreign-owned firms, as the occurrence of these spatial effects is closely linked to the discussion on the need and form of regional development policies in this host economy. Another aspect of the spatiality of the impact of foreign-owned firms that needs to be addressed in future research concerns the difference in externality impact of FDI firms in the border states and in Mexico City, as this difference suggests that FDI firms have played an important role in the locational changes of economic activity in Mexico in the last two decades. Not only will further research improve our understanding of the extent and nature of the differential impact of FDI in these regions, it will also provide us with better insight into how the regional context can affect FDI productivity effects.

Third, more research is required on contemporary effects of FDI firms and firms that participate in the maquiladora programme. The evidence presented in this book contains important corroborating evidence for the notion that the Mexican manufacturing sector contains maquiladora firms of different generations, which vary in their local impact. For instance, the findings in Chapter 6 indicate that there are additional positive FDI externalities in the maquiladora-intensive border states. Also, the findings for Nuevo León suggest that second and third generation maquiladora firms are integrated remarkably well into their local economy, leading to positive externality effects. In contrast, first generation maquiladora firms use significantly less local suppliers and are also significantly less supportive. Of course, the level and nature of FDI backward linkages should be investigated by means of especially designed firm level surveys in regions other than Nuevo León. In such research, it is important to ensure that both foreign-owned and domestic firms are included. The majority of research into FDI local linkages does not include domestic firms, assuming that FDI firms are very likely to use suppliers to a lower degree than domestic firms. The research on producer firms in Nuevo León indicates that such an assumption may be wrong, as the findings in Chapter 7 indicate that in this region there are no significant differences between FDI and Mexican producer firms regarding the level of use of suppliers of both material inputs and production services. Additional research in other regions will not only help to assess the extent to which the findings for Nuevo León can be generalised, research on inter-firm linkages in other Mexican states may also shed additional light on the occurrence of positive and negative externalities through such linkages in different regional settings. In this context, it is also very important that local suppliers are included in future research on externality effects through backward linkages between producer and supplier firms.

Turning to the general research field on FDI externalities, the large extent of heterogeneity of the body of empirical evidence as described in Chapter 2 reflects the importance of careful estimation of FDI effects. As the empirical findings in this book indicate, FDI firms can be linked to positive and negative FDI externalities, both within and between industries. The survey in Chapter 2 shows that most studies only estimate for the presence of intra-industry externalities, thereby omitting potentially important inter-industry effects. Furthermore, as I argue in this book, a likely explanation for the materialisation of negative externalities between FDI and domestic suppliers is that suppliers may be forced to offer concessions to such an extent that their supply relations with FDI firms result in the net creation of negative externalities. More work is needed to improve our understanding of this type of externality effect. In a similar fashion, the effect of the technology gap on FDI externalities needs further clarification. The findings presented in this book, supported by empirical evidence for other host economies, indicate that a large technology gap stimulates positive FDI externalities. It is important to point out that this finding does not dispute the notion that a sufficient level of absorptive capacity among domestic firms is a crucial factor for externalities to materialise. Instead, it suggests that we need to develop alternative indicators of the level of absorptive capacity and try to improve our understanding of the role of technological differences in externality generating and transmitting processes.

Next, I believe it is unfortunate that current empirical research pays relatively little attention to FDI externality effects in developing countries. An important reason for this is the lack of plant level data for many of these countries. In contrast to this, the rapid growth of research on FDI firms in European countries for instance reflects the growing availability of plant level datasets for these countries. However, especially given the central importance of FDI firms for the success of processes of economic liberalisation and trade promotion in many developing countries, it is vital that we obtain more evidence on the impact of FDI in these countries. This may involve decisions on the nature of datasets that can be used, as it seems that cross-sectional data may be more readily available. Although research with this type of data is challenging, this should not be a reason not to use such data to obtain more evidence on the importance of FDI firms in these developing countries. As the research in this book shows, the core criticism of cross-sectional studies that estimates of FDI externalities are always biased upwards is flawed. As the findings for Mexico show, foreign-owned firms may gravitate towards low productivity, labour intensive industries, causing a downward bias in the estimated FDI effect. Of course, similar to the argument that we should not automatically assume that the estimated effect of industry foreign participation will always be biased upwards, the findings for Mexico should not be interpreted as evidence that the opposite will always be the case. Having said so, it seems plausible to assume that a tendency to concentrate in labour intensive industries is likely to apply in particular to cases where FDI firms invest in developing countries to benefit from low labour costs. For these cases, it appears that the instrument that I introduce in this book can be applied to obtain unbiased estimates of FDI externalities. In any case, the findings for Mexico indicate that studies on FDI effects need to pay more attention to the type of bias that may occur in estimating FDI externalities. This applies to both cross-sectional and panel data studies, as it is often not clear to what extent panel data studies actually control successfully for the potential problem of endogeneity that can influence the estimated effect of industry foreign participation. Future research must continue to try and understand whether, how and to what extent estimated effects from foreign participation may be affected by this problem.

Furthermore, I would like to make a case for the continuation and intensification of applied fieldwork, in the form of case studies and especially designed firm level surveys. One reason for this is that often other types of data, especially in developing countries, are simply not available. Also, it is clear that we need to develop a better understanding of why FDI externalities are generated and how they are transmitted. The increasing availability of plant level datasets has generated a large amount of important empirical evidence. However, most of these datasets usually contain little additional information besides variables indicating inputs, output, type of ownership and perhaps location. With this type of data, it is difficult to obtain a better understanding of when and how FDI externalities occur, let alone to identify the individual contributions of the various channels that transmit these effects. Case studies and especially designed firm level surveys are time consuming, expensive and need to be carried out with great care, but if done correctly, they can improve our understanding of FDI externality effects substantially.

Finally, the empirical findings presented in this book give strong support to the notion that spatial dimensions of FDI externalities can be very important. Research on the location patterns of industries and the generation and transmission of effects from knowledge creation assign great importance to concepts of agglomeration, geographical proximity, knowledge spillovers and geographical distance. At the same time, there is a clear need in applied research on FDI effects to identify new determinants of FDI externalities. In combination, this strongly suggests that future research on FDI effects must include these important geographical dimensions. The empirical studies that do so, in combination with the empirical findings presented in this book, represent initial attempts to identify and estimate the role of spatial dimensions in processes that generate and transmit FDI externalities. As such, they form an important starting point from which future research can be developed. In fact, just as these spatial dimensions play a key role in contemporary studies on externalities related to R&D and other types of knowledge creating activities, I see no reason why the incorporation of these spatial dimensions should not be given the same central role in future applied studies on FDI effects.

As the findings in this book suggest, there are several aspects of the relation between FDI externalities and geographical space which future research needs to address. One important issue is that more research on the relation between agglomeration and FDI effects is urgently called for. We know from empirical studies on FDI location factors that FDI firms tend to gravitate to regions within host economies that contain agglomerations of economic activity. Also, the high degree of similarity between the channels of FDI externalities and the mechanisms that underlie agglomeration economies strongly suggest that agglomeration will affect these FDI effects. The fact that this relation has been addressed in only a very limited number of studies indicates the need for more evidence on this relation. Also, the need for further research is underlined by the findings in this book that suggest that agglomeration can promote both positive and negative FDI externalities. Given these findings, we need to obtain more evidence and a better understanding of how exactly this relation between agglomeration and FDI spillovers can materialise. With respect to the issue whether agglomeration promotes positive or negative externalities, the findings for Mexico also suggest that the regional context is likely to be an important element of the relationship between agglomeration and FDI externalities, suggesting that other regional characteristics besides agglomeration or geographical proximity may influence these FDI effects. A modest piece of evidence that supports this idea is represented by the findings in Chapter 6 that indicate that regional specialisation and diversity both appear to influence FDI externalities between Mexican regions. This suggests that future research on FDI effects may benefit from analysing the relation between FDI effects and a variety of regional characteristics besides agglomeration, in the continuing search for better evidence and a clearer understanding of factors that influence FDI productivity effects.

The second main issue that the findings for Mexico relate to is that more work is needed on the relation between geographical distance and FDI externalities. Research on spatial FDI spillovers has remained limited in scale thus far, and most research strategies and specifications are ad-hoc and incomplete in many cases. In this book, I provide a modest contribution in this respect. For instance, I distinguish between FDI effects at the national and the regional level, which turns out to be very important. Furthermore, I experiment with a variety of specifications of the relation between geographical distance and FDI externalities and I introduce improved indicators of the scope of potential spatial FDI effects. In addition to confirming that there do appear to be spatial FDI effects between Mexican regions, the findings also indicate that the results are sensitive to the specification of the relation between distance and externalities and to how the scope of potential spatial effects is measured. Therefore, although the findings confirm that spatial dimensions are important, they also reflect the scope for future research to further investigate the role of geographical dimensions in the generation and transmission of FDI externalities.

References

Abraham, F., Konings, J. and Slootmaekers, V. 2007. FDI spillovers in the Chinese manufacturing sector: Evidence of firm heterogeneity. *CEPR Discussion Paper Series*, no. 6573. London: Centre for Economic Performance.

Abramovitz, M. 1986. Catching up, forging ahead and falling behind. *The Journal of Economic History*, 46(2), 385-406.

Abreu, M., de Groot, H.LO.F. and Florax, R.J.G.M. 2004. Space and growth. *Tinbergen Institute Discussion Papers Series*, no. 129/3. Amsterdam: Tinbergen Institute.

Acs, Z.J., Audretsch, D.B. and Feldman, M.P. 1994. R&D spillovers and recipient firm size. *Review of Economics and Statistics*, 76(2), 336-340.

Adserá, A. 2000. Sectoral spillovers and the price of land: A cost analysis. *Regional Science and Urban Economics*, 30(5), 565-585.

Aguilar, A. 1999. Mexico City growth and regional dispersal: the expansion of largest cities and new spatial forms. *Habitat International*, 23(3), 391-412.

Aitken, B., Harrison, A. and Lipsey, R.E. 1996. Wages and foreign ownership: A comparative study of Mexico, Venezuela and the United States. *Journal of International Economics*, 40(3-4), 345-371.

Aitken, B., Hanson, G. and Harrison, A.E. 1997. Spillovers, foreign investment and export behaviour. *Journal of International Economics*, 43(1-2), 103-132.

Aitken, B.J. and Harrison, A.E. 1999. Do domestic firms benefit from direct foreign investment? Evidence from Venezuela. *The American Economic Review*, 89(3), 605-618.

Alba, F. 1982. *The population of Mexico: Trends, issues and policies*. New Brunswick, N.J.: Transaction Books.

Alfaro, L., Chanda, A. and Kalemli-Ozcan, S. 2004. FDI and economic growth: The role of financial markets. *Journal of International Economics*, 64(1), 89-112.

Alfaro, L. and Rodríguez-Clare, A. 2004. Multinationals and linkages: An empirical investigation. *Economía*, Spring issue, 113-169.

Alonso-Villar, O. 1999. Spatial distribution of production and international trade: A note. *Regional Science and Urban Economics*, 29(3), 371-380.

Altenburg, T., Bosse, D., Brunzema, T., Eckhardt, J., Unger, B. and Zeeb, S. 1998. *Entwicklung und förderung von zulieferindustrien in Mexico*. Berlin: Deutsches Institut für Entwicklungspolitiek.

Angrist, J.D. and Krueger, A.B. 1991. Does compulsory school attendance affect schooling and earnings? *Quarterly Journal of Economics*, 106(4), 979-1014.

Anselin, L. 1988. *Spatial econometrics: Methods and models*. Dordrecht: Kluwer.

Anselin, L., Acs, Z. and Varga, A. 1997. Local geographic spillovers between university research and high technology innovations. *Journal of Urban Economics*, 42(3), 422-448.

Anselin, L., Varga, A. and Acs, Z. 2000. Geographic and sectoral characteristics of academic knowledge externalities. *Papers in Regional Science*, 79(4), 435-443.

Aroca, P., Bosch, M. and Maloney, W.F. 2005. Spatial dimensions of trade liberalization and economic convergence: Mexico 1985-2002. *World Bank Economic Review*, 19(3), 345-378.

Audretsch, D.B. and Feldman, M.P. 2004. Knowledge spillovers and the geography of innovation, in *Handbook of urban and regional economics, vol. 4*, edited by Henderson, V.J. and Thisse, J.F. Amsterdam: North Holland.

Balasubramanyam, V.N., Salisu, M. and Sapsford, D. 1996. Foreign direct investment and growth in EP and IS countries. *The Economic Journal*, 106(434), 92-105.

Baldwin, R., Forslid, R., Martin, P., Ottaviano, G. and Robert-Nicoud, F. 2003. *Economic geography and public policy*. Princeton: Princeton University Press.

Barrios, S. 2000. Foreign direct investment and productivity spillovers: Evidence from the Spanish experience. *Documento de Trabajo*. Madrid: Fundación de Estudios de Economía Aplicada.

Barrios, S. and Strobl, E. 2002. Foreign direct investment and productivity spillovers: Evidence from the Spanish experience. *Review of World Economics*, 138(3), p. 459-481.

Barrios, S., Bertinelli, L. and Strobl, E. 2006. Co-agglomeration and spillovers. *Regional Science and Urban Economics*, 36(4), 467-481.

Barrow, R.J. and Sala-i-Martin, X. 1995. *Economic growth*. Cambridge, MA: The MIT Press.

Barry, F., Görg, H. and Strobl, E. 2003. Foreign direct investment, agglomerations, and demonstration effects: An empirical investigation. *Review of World Economics*, 139(4), 583-600.

Barry, F., Görg, H. and Strobl, E. 2005. Foreign direct investment and wages in domestic firms in Ireland: Productivity spillovers versus labour-market crowding out. *International Journal of the Economics of Business*, 12(1), 67-84.

Bartik, T.J. 1985. Business location decisions in the United States: Estimates of the effects of unionisation, taxes, and other characteristics of states. *Journal of Business and Economic Statistics*, 3(1), 14-22.

Basile, R., Castellani, D. and Zanfei, A. 2003. Location choices of multinational firms in Europe: The role of national boundaries and EU policy. *Working Paper Series on Economics, Mathematics and Statistic*, no. 78. University of Urbino, Italy.

Basu, P., Chakraborty, C. and Reagle, D. 2003. Liberalisation, FDI, and growth in developing countries: A panel cointegration approach. *Economic Enquiry*, 41(3), 510-516.

Bator, F.M. 1958. The anatomy of market failure. *Quarterly Journal of Economics*, 72, 351-379.

Békés, G., Kleinert, J. and Toubal, F. 2006. Spillovers from multinationals to heterogeneous domestic firms: Evidence from Hungary. *Discussion Paper Series*, no. 2006/16, Institute of Economics, Hungarian Academy of Sciences.

Belderbos, R., Capannelli, G. and Fukao, K. 2001. Backward vertical linkages of foreign manufacturing affiliates: Evidence from Japanese multinationals. *World Development*, 29(1), 189-208.

Bethell. L. 1984. *The Cambridge History of Latin America*. Cambridge: Cambridge University Press.

Biles, J.J. 2004. Export-oriented industrialization and regional development: A case study of maquiladora production in Yucatán, Mexico. *Regional Studies*, 38(5), 519-534.

Blalock, G. and Gertler, P.J. 2008. Welfare gains from foreign direct investment through technology transfer to local suppliers. *Journal of International Economics*, 74(2), 402-421.

Blomström, M. 1986. Foreign investment and productive efficiency: The case of Mexico. *Journal of Industrial Economics*, 35(1), 97-110.

Blomström, M. 1989. *Foreign Investment and spillovers*. London: Routledge.

Blomström, M. and Persson, H. 1983. Foreign investment and spillover efficiency in an underdeveloped economy: Evidence from the Mexican manufacturing industry. *World Development*, 11(6), 493-501.

Blomström, M. and Wang, J.-Y. 1992. Foreign investment and technology transfer: A simple model. *European Economic Review*, 36(1), 137-155.

Blomström, M. and Wolff, E. 1994. Multinational corporations and productivity convergence in Mexico, in *Convergence of productivity: Cross-national studies and historical evidence*, edited by Baumol, W., Nelson, R. and Wolff, E. Oxford: Oxford University Press.

Blomström, M. and Kokko, A. 1997. Regional integration and foreign direct investment. *NBER Working Papers*, no. 6019. Cambridge, MA: National Bureau of Economic Research.

Blomström, M. and Kokko, A. 1998. Multinational corporations and spillovers. *Journal of Economic Surveys*, 12(3), 1-31.

Blomström. M. and Sjöholm, F. 1999. Technology transfer and spillovers: Does local participation with multinationals matter? *European Economic Review*, 43(4-6), 915-932.

Blomström, M. and Kokko, A. 2003. The economics of foreign direct investment incentives. *NBER Working Paper Series*, no. 9489. Cambridge, MA: National Bureau of Economic Research.

Blomström, M., Globerman, S. and Kokko, A. 1999. The determinants of host country spillovers from foreign direct investment: A review and synthesis of the literature. *Working Paper Series in Economics and Finance*, no. 399. Stockholm: Stockholm School of Economics.

Blomström, M., Lipsey, R.E. and Zejan, M. 1994. What explains the growth of developing countries?, in *Convergence of productivity: Cross-national studies and historical evidence*, edited by Baumol, W., Nelson, R. and Wolff, E. Oxford: Oxford University Press.

Blomström, M., Kokko, A. and Zejan, M. 2000. *Foreign direct investment: Firm and host economy strategies*. Basingstoke: Macmillan.

Blyde, J., Kugler, M. and Stein, E. 2004. Exporting versus outsourcing by MNC subsidiaries: Which determines FDI spillovers? *Discussion Papers in Economics and Econometrics*. Southampton: University of Southampton.

Boarnet, M.G. 1998. Spillovers and the locational effects of public infrastructure. *Journal of Regional Science*, 38(3), 381-400.

Bode, E. 2004. The spatial pattern of localised R&D spillovers: An empirical investigation for Germany. *Journal of Economic Geography*, 4(1), 43-64.

Borensztein, E., De Gregorio, J. and Lee, J.W. 1998. How does foreign direct investment affect economic growth? *Journal of International Economics*, 45(1), 115-135.

Bound, J., Jaeger, D.A. and Baker, R. 1993. The cure can be worse than the disease: A cautionary tale regarding instrumental variables. *NBER Technical Paper Series*, no. 137. Cambridge, MA: National Bureau of Economic Research.

Brakman, S., Garretsen, H. and van Marrewijk, C. 2001. *An introduction to geographical economics*. Cambridge: Cambridge University Press.

Brannon, J.T., James, D.D. and Lucker, G.W. 1994. Generating and sustaining backward linkages between maquiladoras and local suppliers in northern Mexico. *World Development*, 22(12), 1933-1945.

Buitelaar, R.M. and Padilla Perez, R. 2000. Maquila, economic reform and corporate strategies. *World Development*, 28(9), 1627-1642.

Canfei He 2002. Information costs, agglomeration economies and the location of foreign direct investment in China. *Regional Studies*, 36(9), 1029-1036.

Canfei He 2003. Location of foreign manufacturers in China: Agglomeration economies and country of origin effects. *Papers in Regional Science*, 82(3), 351-372.

Cárdenas, E. 1996. *La Política Económica en México: 1950-1994*. México, D.F.: Fondo de Cultural Económica. El Colegio de México.

Carkovic, M. and Levine, R. 2005. Does foreign direct investment accelerate economic growth? in *Does foreign direct investment promote development?*, edited by Moran, T.H., Graham, E.M. and Blomström, M. Washington, DC: Institute for International Economics.

Carlton, D. 1983. The location and employment choices of new firms: An econometric model with discrete and continuous variables. *Review of Economics and Statistics*, 65(3), 440-449.

Carrillo, J. 1995. Flexible production in the auto sector: Industrial reorganisation at Ford-Mexico. *World Development*, 23(1), 87-101.

Carrillo, J. and Mortimore, M. 1998. Competividad de la industria de televisores en México. *Revista Latinoamericana de Estudios do Trabalho*, vol. 4.

Carrillo, J. and Hualde, A. 1998. Third generation maquiladora? The Delphi-General Motors case. *Journal of Borderland Studies*, 13(1), 79-97.

Carrillo, J. and Gomis, R. 2003. Challenge for maquiladoras given the loss of competitiveness. *Comercio Exterior*, vol. 53.

Castellani, D. and Zanfei, A. 2003. Productivity gaps, inward investments and productivity of European firms. *Economics of Innovation and New Technology*, 12, 450-468.

Caves, R.E. 1974. Multinational firms, competition and productivity in host-economy markets. *Economia*, 41(162), 176-193.

Caves, R.E. 1996. *Multinational enterprise and economic analysis*. Cambridge: Cambridge University Press.

CEPAL 1996. *México: La industria maquiladora*. Santiago, Chile: Comisión Económica para America Latina.

Chiquiar, D. 2005. Why Mexico's regional income divergence broke down. *Journal of Development Economics*, 77(1), 257-275.

Chiquiar, D. 2008. Globalisation, regional wage differentials and the Storper-Samuelson theorem: Evidence from Mexico. *Journal of International Economics*, 74(1), 70-93.

Chowdury, A. and Mavrotas, G. 2006. FDI and growth: What causes what? *The World Economy*, 29(1), 9-19.

Chuang, Y.-C. and Lin, C.-M. 1999. Foreign direct investment, R&D and spillover efficiency: Evidence from Taiwan's manufacturing firms. *Journal of Development Studies*, 35(4), 117-137.

Ciccone, A. and Hall, R.E. 1996. Productivity and the density of economic activity. *The American Economic Review*, 86(1), 54-70.

Cohen, W.M. and Levinthal, D.A. 1989. Innovation and learning: The two faces of R&D. *The Economic Journal*, 99(397), 569-596.

Cohen W.M. and Levinthal, D.A. 1990. Absorptive capacity: A new perspective on learning and innovation. *Administrative Science Quarterly*, 35, 128-152.

Cole, E.T. and Ensign, P.C. 2005. An examination of US FDI into Mexico and its relation to NAFTA. *The International Trade Journal*, 19, 1-30.

Combes, P.-P. 2000. Economic structure and local growth: France 1984-1993. *Journal of Urban Economics*, 47(3), 329-355.

Combes, P.-P. and Overman, H.G.O. 2004. The spatial distribution of economic activities in the European Union, in *Handbook of urban and regional economics, vol. 4*, edited by Henderson, V.J. and Thisse, J.F. Amsterdam: North Holland.

Cooney, P. 2001. The Mexican crisis and the maquiladora boom: A paradox of development or the logic of neoliberalism? *Latin American Perspectives*, 28(3), 55-83.

Coughlin, C. and Segev, E. 2000. Location determinants of new foreign-owned manufacturing plants. *Journal of Regional Science*, 40(2), 323-351.

Coughlin, C., Terza, J.V. and Arromdee, V. 1991. State characteristics and the location of foreign direct investment in the United States. *The Review of Economics and Statistics*, 73(4), 675-678.

Crescenzi, R., Rodríguez-Pose, A. and Storper, M. 2007. The territorial dynamics of innovation: A Europe–United States comparative analysis. *Journal of Economic Geography*, 7(6), 673-709.

Crespo, N. and Fontoura, M.O. 2007. Determinant factors of FDI spillovers – What do we really know? *World Development*, 35(3), 410-425.

Crespo, N., Proença, I. and Fontoura, M.P. 2007. FDI spillovers at the regional level: Evidence from Portugal. *School of Economics and Management Working Paper Series*, no. 28. Lisbon: Technical University of Lisbon.

Crozet, M. and Koenig Soubeyran, P. 2004. EU enlargement and the internal geography of countries. *Journal of Comparative Economics*, 32(2), 265-279.

Crozet, M., Mayer, T. and Muchielli, J.-L. 2004. How do firms agglomerate? A study of FDI in France. *Regional Science and Urban Economics*, 34(1), 27-54.

Cuadros, A., Orts, V. and Alguacil, M. 2004. Openness and growth: Re-examining foreign direct investment, trade and output in Latin America. *Journal of Development Studies*, 40(4), 167-192.

Cuevas, A., Messmacher, M. and Werner, A. 2005. Foreign direct investment in Mexico since the approval of NAFTA. *World Bank Economic Review*, 19(3), 473-488.

Cypher, J. 2001. NAFTA's lessons: From economic mythology to current realities. *Labour Studies Journal*, 26(1), 5-21.

Damijan, J.P., Knell, M., Macjen, B. and Rojec, M. 2003a. The role of FDI, R&D accumulation, and trade in transferring technology to transition countries: Evidence from firm panel data for eight transition countries. *Economic Systems*, 27(2), 189-204.

Damijan, J.P., Knell, M., Macjen, B. and Rojec, M. 2003b. Technology transfer through FDI in top-10 transition countries: How important are direct effects, horizontal and vertical spillovers? *William Davidson Institute Working Paper Series*, no. 549. Michigan: William Davidson Institute, University of Michigan.

De Mello, L. 1997. Foreign direct investment in developing countries and growth: A selective survey. *Journal of Development Studies*, 34, 1-34.

De Mello, L. 1999. Foreign direct investment-led growth: Evidence from time series and panel data. *Oxford Economic Papers*, 51(1), 133-151.

De Propis, L. and Driffield, N. 2006. The importance of clusters for spillovers from foreign direct investment and technology sourcing. *Cambridge Journal of Economics*, 30(2), 277-291.

Dimelis, S. and Louri, H. 2002. Foreign ownership and production efficiency: A quantile regression analysis. *Oxford Economic Papers*, 54(3), 449-469.

Djankov, S. and Hoekman, B. 2000. Foreign investment and productivity growth in Czech enterprises. *World Bank Economic Review*, 14(1), 49-64.

Döring, T. and Schellenbach, J. 2006. What do we know about geographical knowledge spillovers and regional growth? A survey of the literature. *Regional Studies*, 40(3), 375-395.

Driffield, N. 2001. The impact of domestic productivity on inward investment in the UK. *Manchester School*, 69(1), 103-119.

Driffield, N. 2004. Regional policy and spillovers from FDI in the UK. *Annals of Regional Science*, 38(4), 579-594.

Driffield, N. 2006. On the search for spillovers from FDI with spatial dependency. *Regional Studies*, 40(1), 107-119.

Driffield, N.L. and Noor, A.H.M. 2000. Foreign direct investment and local input linkages in Malaysia. *Transnational Corporations*, 8(3), 1-25.

Driffield, M. and Munday, M. 2000. Industrial performance, agglomeration, and foreign manufacturing investment in the UK. *Journal of International Business Studies*, 31(1), 21-37.

Driffield, M. and Munday, M. 2001. Foreign manufacturing, regional agglomeration and technical efficiency in UK manufacturing industries: A stochastic production frontier approach. *Regional Studies*, 35(5), 391-399.

Driffield, N. and Girma. S. 2003. Regional foreign direct investment and wage spillovers: Plant level evidence from the UK electronics industry. *Oxford Bulletin of Economics and Statistics*, 65(4), 453-474.

Driffield, N. and Love, J.H. 2007. Linking FDI motivation and host economy productivity effects: Conceptual and empirical analysis. *Journal of International Business Studies*, 38(3), 460-473.

Driffield, N., Munday, M. and Roberts, A. 2004. Inward investment, transaction linkages and productivity spillovers. *Papers in Regional Science*, 83(4), 699-722.

Drukker, D.M. 2003. Testing for serial correlation in linear panel-data models. *Stata Journal*, 3(2), 168-177.

Dunning, J.H. 1993. *Multinational enterprises and the global economy*. Reading, MA: Addison-Wesley Publishing Company.

Duranton, G. and Puga, D. 2004. Micro-foundations of urban agglomeration economies, in *Handbook of urban and regional economics, vol. 4*, edited by Henderson, V.J. and Thisse, J.F. Amsterdam: North Holland.

Durham, J.B. 2004. Absorptive capacity and the effects of foreign direct investment and equity foreign portfolio investment on economic growth. *European Economic Review*, 48(2), 285-306.

Dussel, P. 1999. La subcontratación como proceso de aprendizaje: El caso de la electrónica en Jalisco (México) en la década de los noventa. *CEPAL Series Desarrollo Productivo*, no. 55. Santiago, Chile: Comisión Económica para America Latina.

Eberts, R.W. and McMillen, D. 1999. Agglomeration economies and public infrastructure, in *Handbook of urban and regional economics, vol. 3*, edited by Cheshire, P. and Mills, E.S. Amsterdam: North Holland.

ECLAC 1999. *Foreign investment in Latin America and the Caribbean*. Santiago, Chile: Economic Commission for Latin America and the Caribbean.

Esquivel, G. and Messmacher, M. 2002. *Sources of regional (non) convergence in Mexico*. Washington DC: IBRD Mimeo, Chief Economist office for Latin America.

Ewe-Ghee Lim 2001. Determinants of, and the relation between, foreign direct investment and growth: A summary of the recent literature. *IMF Working Paper Series*, no. 01/175. Washington, DC: International Monetary Fund.

Faber, B. 2007. Towards the spatial patterns of sectoral adjustments to trade liberalisation: The case of NAFTA in Mexico. *Growth and Change*, 38(4), 567-595.

Feenstra, R.C. and Hanson, G. 1997. Foreign direct investment and relative wages: Evidence from Mexico's maquiladoras. *Journal of International Economics*, 42(3-4), 371-393.

Findlay, R. 1978. Relative backwardness, direct foreign investment and the transfer of technology: A simple dynamic model. *The Quarterly Journal of Economics*, 92(1), 1-16.

Fosfuri, A., Motta, M. and Ronde, T. 2001. Foreign direct investment and spillovers through workers' mobility. *Journal of International Economics*, 53(1), 205-222.

Friedman, J., Gerlowski, D.A. and Silberman, J. 1992. What attracts foreign multinational corporations? Evidence from branch plant location in the United States. *Journal of Regional Science*, 32(4), 403-418.

Friedman, J., Hung-Gay Fung, Gerlowski, D.A. and Silberman, J. 1996. A note on 'State characteristics and the location choice of foreign direct investment within the United States'. *The Review of Economics and Statistics*, 78(2), 367-368.

Fuentes, A.N., Alegría, T., Brannon, J., Dilmus, J. and Luckner, G.W. 1993. Local sourcing and indirect employment: Multinational enterprises in Northern Mexico. In, Bailey, P., Parisotto, A. and Renshaw, G. (eds) *Multinationals and employment. The global economy of the 1990s*. Geneva: International Labour Office.

Fujita, M., Krugman, P. and Venables, A.J. 1999. *The spatial economy: Cities, regions and international trade*. Cambridge, MA: The MIT Press.

Gallagher, K.P. and Zarsky, L. 2004. Sustainable industrial development? The performance of Mexico's FDI-led integration strategy. *Global Development and Environment Institute*, Fletcher School of Law and Diplomacy, Tuft University.

Garza, G. 1999. Global economy, metropolitan dynamics and urban policies in Mexico. *Cities*, 16, 149-170.

Gershenkron, A. 1962. *Economic backwardness in historical perspective*. Cambridge, MA: Belknap Press of Harvard University Press.

Girma, S. 2005. Absorptive capacity and productivity spillovers from FDI: A threshold regression analysis. *Oxford Bulletin of Economics and Statistics*, 67(3), 281-306.

Girma, S. and Görg, H. 2007. The role of the efficiency gap for spillovers from FDI: Evidence from the UK electronics and engineering sectors. *Open Economies Review*, 18(2), 215-232.

Girma, S. and Wakelin, K. 2007. Local productivity spillovers from foreign direct investment in the UK electronics industry. *Regional Science and Urban Economics*, 37(3), 399-412.

Girma, S., Greenaway, D. and Wakelin, K, 2001. Who benefits from foreign direct investment in the UK? *Scottish Journal of Political Economy*, 48(2), 239-248.

Girma, S., Görg, H. and Pisu, M. 2008. Exporting, linkages and productivity spillovers from foreign direct investment. *Canadian Journal of Economics*, 41(1), 320-340.

Giroud, A. and Hafiz Mirza 2006. Factors determining supply linkages between transnational corporations and local suppliers in ASEAN. *Transnational Corporations*, 15(3), 1-34.

Glaeser, E.L., Hedi, D.K., Scheinkman, J.A. and Schleifer, A. 1992. Growth in cities. *The Journal of Political Economy*, 100(6), 143-156.

Globerman, S. 1979. Foreign direct investment and spillover efficiency in Canadian manufacturing industries. *Canadian Journal of Economics*, XX(1), 42-56.

Gordon, I.R. and McCann, P. 2000. Industrial clusters: Complexes, agglomeration and/or social networks? *Urban Studies*, 37(3), 513-532.

Görg, H. and Strobl, E. 2001. Multinational companies and productivity spillovers: A meta-analysis. *The Economic Journal*, 111(475), 723-739.

Görg, H. and Greenaway, D. 2004. Much ado about nothing? Do domestic firms really benefit from foreign direct investment? *World Bank Research Observer*, 19(2), 171-197.

Graham, E. and Wada, E. 2000. Domestic reform, trade and investment liberalisation, financial crisis and foreign direct investment into Mexico. *World Economy*, 23(6), 777-797.

Grether, J.M. 1999. Determinants of technology diffusion in Mexican manufacturing: A plant-level analysis. *World Development*, 27(7), 1287-1298.

Grilliches, Z. 1979. Issues in assessing the contribution of R&D to productivity growth. *Bell Journal of Economics*, 10(1), 92-116.

Grilliches, Z. 1992. The search for R&D spillovers. *Scandinavian Journal of Economics*, 94, 29-47.

Grossman, G. and Helpman, E. 1994. Endogenous innovation in the theory of growth. *Journal of Economic Perspectives*, 8(1), 23-44.

Guimarães, P., Figueiredo, O. and Woodward, D. 2000. Agglomeration and the location of foreign direct investment in Portugal. *Journal of Urban Economics*, 47(1), 115-135.

Haddad, M. and Harrison, A. 1993. Are there positive spillovers from direct foreign investment? Evidence form panel data for Morocco. *Journal of Development Economics*, 42(1), 51-74.

Hallbach, A.J. 1989. Multinational enterprises and subcontracting in the third world: A study of inter-industry linkages. *ILO Working Paper Series*, no 58. Geneva: International Labour Office.

Halpern, L. and Muraközy, B. 2007. Does distance matter in spillover? *Economics of Transition*, 15, 781-805.

Hanson, G. 1996. Localisation economies, vertical organisation, and trade. *The American Economic Review*, 86(5), 1266-1278.

Hanson, G. 1997. Increasing returns, trade, and the regional structure of wages. *The Economic Journal*, 107(440), 113-133.

Hanson, G. 1998a. North American economic integration and industry location. *Oxford Review of Economic Policy*, 14(2), 30-44.

Hanson, G. 1998b. Regional adjustment to trade liberalisation. *Regional Science and Urban Economics*, 28(4), 419-444.

Hanson, G. 2001a. Scale economies and the geographic concentration of industry. *Journal of Economic Geography*, 1(3), 255-276.

Hanson, G. 2001b. Should countries promote foreign direct investment? *G-24 discussion paper series*, no. 9. Centre for International Development, Harvard University, Cambridge, M.A.

Harris, C.D. 1954. The market as a factor in the localisation of industry in the United States. *Annals of the Association of American Geographers*, 44, 315-348.

Haskel, J.E., Pereira, S.C. and Slaughter, M.J. 2007. Does inward foreign direct investment boost the productivity of domestic firms? *The Review of Economics and Statistics*, 89(3), 482-496.

Hausman, J. 1978. Specification tests in econometrics. *Econometrica*, 46(6), 1251-1271.

Head, K.C., Ries, J.C. and Swenson, D.L. 1995. Agglomeration benefits and location choice: Evidence from Japanese manufacturing investments in the United States. *Journal of International Economics*, 38(3-4), 223-247.

Head, K.C., Ries, J.C. and Swenson, D.L. 1999. Attracting foreign manufacturing: Investment promotion and agglomeration. *Regional Science and Urban Economics*, 29(2), 197-218.

Head, K.C. and Mayer, T. 2004. Market potential and the location of Japanese investment in the European Union. *The Review of Economics and Statistics*, 86(4), 959-972.

Henderson, V.J. 1988. *Urban development: Theory, fact and illusion*. Oxford: Oxford University Press.

Henderson, V.J. 1997. Externalities and industrial development. *Journal of Urban Economics*, 42(3), 449-470.

Henderson, V.J. 2001. Urban scale economies. In: Paddison, R. (ed.) *Handbook of urban studies*. London: Sage.

Henderson, V.J. 2003. Marshall's scale economies. *Journal of Urban Economics*, 53(1), 1-28.

Henderson, V.J. 2007. Understanding knowledge spillovers. *Regional Science and Urban Economics*, 37(4), 497-508.

Henderson, V.J., Kuncoro, A. and Turner, M. 1995. Industrial development in cities. *The Journal of Political Economy*, 103(5), 1067-1090.

Henderson, V.J., Shalizi, Z. and Venables, A.J. 2001. Geography and development. *Journal of Economic Geography*, 1(1), 81-105.

Hermes, N. and Lensink, R. 2003. Foreign direct investment, financial development and economic growth. *Journal of Development Studies*, 40(1), 142-163.

Hilber, C.A.L. and Voicu, I. 2005. Agglomeration economies and the location of foreign direct investment. Quasi-experimental evidence from Romania. *Forthcoming in Regional Studies*.

Imbriani, C. and Reganiti, F. 1997. Spillovers internazionali di eficienza net settore manifatturiero italiano. *Economia Internazionale*, 50, 583-595.

Intrilligator, M.D., Bodkin, R.G. and Hsiao, C. 1996. *Econometric models, techniques, and applications*. London: Prentice-Hall International (UK) Limited.

Ivarsson, I. and Alvstam, C.G. 2005. Technology transfer from TNCs to local suppliers in developing countries: A study of AB Volvo's truck and bus plants in Brazil, China, India and Mexico. *World Development*, 33(8), 1325-1344.

Jaffe, A.B. 1989. Real effects of academic research. *The American Economic Review*, 79(5), 957-970.

Jaffe, A.B., Trajtenberg, M. and Henderson, R. 1993. Geographic localisation of knowledge spillovers as evidenced by patent citations. *Quarterly Journal of Economics*, 108(3), 577-598.

Jaffe, A.B. and Trajtenberg, M. 2002. *Patents, citations and innovations: A window on the knowledge economy*. Cambridge, MA: The MIT Press.

Javorcik, B.S. 2008. Can survey evidence shed light on spillovers from foreign direct investment? *World Bank Research Observer*, 23(2), 139-159.

Javorcik, B.S. and Spatareanu, M. 2003. To share or not to share: Does local participation matter for FDI spillovers? *World Bank Policy Research Working Paper*, no. 3118. Washington, DC: World Bank.

Javorcik, B.S., Saggi, K. and Spatareanu, M. 2004. Does it matter where you come from? Vertical spillovers from foreign direct investment and the nationality of investors. *World Bank Policy Research Working Paper Series*, no. 3449. Washington, DC: World Bank.

Javorcik, B.S. and Spatareanu, M. 2005. Disentangling FDI spillover effects: What do firm perceptions tell us? in, *Does foreign direct investment promote development?* edited by Moran, T.H., Graham, E.M. and Blomström, M. Washington, DC: Institute for International Economics.

Jordaan, J.A. 2004a. *Foreign direct investment, externalities and geography*. Unpublished PhD thesis. London: London School of Economics and Political Science.

Jordaan, J.A. 2004b. Estimating FDI-induced externalities when FDI is endogenous: A comparison between OLS and IV estimates of FDI-induced externalities in Mexico. *Research Papers in Environmental and Spatial Analysis*, no. 93. Department of Geography and Environment. London: London School of Economics and Political Science.

Jordaan, J.A. 2005. Determinants of FDI-induced externalities: New empirical evidence for Mexican manufacturing industries. *World Development*, 33(12), 2103-2118.

Jordaan, J.A. 2008a. State characteristics and the locational choice of foreign direct investment: Evidence from regional FDI in Mexico 1989-2006. *Growth and Change*, 39(3), 389-413.

Jordaan, J.A. 2008b. Regional foreign participation and externalities: New empirical evidence from Mexican regions. *Environment and Planning A*, 40(12), 2948-2969.

Jordaan, J.A. 2008c. Intra- and inter-industry externalities from foreign direct investment in the Mexican manufacturing sector: New evidence from Mexican regions. *World Development*, 36(12), 2838-2854.

Jordaan, J.A. and Harteveld, L. 1997. *Economic impact of multinational enterprises in newly industrialising economies: A case study of the manufacturing sector in Nuevo León, Mexico*. University of Utrecht Research Series. Utrecht: The Netherlands.

Jordaan, J.A. and Sanchez-Reaza, J. 2006. Trade liberalisation and location: Empirical evidence for Mexican manufacturing industries 1980-2003. *The Review of Regional Studies*, 36, 279-303.

Juan Ramón, V.H. and Rivera-Batíz, L.A. 1996. Regional growth in Mexico 1973-1993. *IMF Working Paper Series*, no. WP/96/92. Washington, DC: International Monetary Fund.

Kaldor, N. 1970. The case for regional policies. *Scottish Journal of Political Economy*, 17(3), p. 337-348.

Karpaty, P. and Lundberg, L. 2007. Foreign direct investment and productivity spillovers in Swedish manufacturing. *Working Paper Series*, no. 2004:2. Swedish Business School, Örebro University.

ten Kate, A. 1992. Trade liberalisation and economic stabilization in Mexico: Lessons of experience. *World Development*, 20(5), 659-672.

Kathuria, V. 2001. Technology transfer and knowledge spillovers to Indian manufacturing firms – A stochastic frontier analysis. *Applied Economics*, 33(5), 625-642.

Kathuria, V. 2002. Liberalisation, FDI and productivity spillovers – An analysis of Indian manufacturing firms. *Oxford Economic Papers*, 54(4), 688-718.

Katz, J.M. 1969. *Production functions, foreign investment and growth*. Amsterdam: North Holland.

Keller, W. 1996. Absorptive capacity: On the creation of and acquisition of technology in development. *Journal of Development Economics*, 49(1), 199-227.

Keller, W. 2004. International Technology Diffusion. *Journal of Economic Literature*, 42(3), 752-782.

Keller, W. and Yeaple, S.R. 2003. Multinational enterprises, international trade, and productivity growth: Firm-level evidence from the United States. *NBER Working Paper Series*, no. 9504, Cambridge, MA: National Bureau of Economic Research.

Kenney, M. and Florida, R. 1994. Japanese maquiladoras: Production organization and global commodity chains. *World Development*, 22(1), 27-44.

Kholdy, S. 1995. Causality between foreign investment and spillover efficiency. *Applied Economics*, 27(8), 745-749.

Kinoshita, Y. 2001. R&D and technology spillovers through FDI: Innovation and absorptive capacity. *William Davidson Institute Working Paper Series*, no. 349. Michigan: William Davidson Institute, University of Michigan.

Kokko, A. 1994. Technology, market characteristics and spillovers. *Journal of Development Economics*, 43(2), 279-293.

Kokko, A. 1996. Productivity spillovers from competition between local firms and foreign affiliates. *Journal of International Development*, 8(4), 517-530.

Kokko, A., Tansini, R. and Zejan, M. 1996. Local technological capability and productivity spillovers from FDI in the Uruguayan manufacturing sector. *Journal of Development Studies*, 32(4), 602-611.

Kokko, A., Zejan, M. and Tansini, R. 2001. Trade regimes and spillover effects from FDI: Evidence from Uruguay. *Review of World Economics*, 137(1), 124-149.

Konings, J. 2000. The effects of foreign direct investment on domestic firms: Evidence from firm level panel data in emerging economics. *William Davidson Institute Working Paper Series*, no. 344. Michigan: William Davidson Institute, University of Michigan.

Krugman, P. 1991. *Geography and trade*. Cambridge, MA: The MIT Press.

Krugman, P. and Venables, A.J. 1995. Globalisation and the inequality of nations. *Quarterly Journal of Economics*, 60(4), 857-880.

Krugman, P. and Elizondo-Livas, R. 1996. Trade policy and the third world metropolis. *Journal of Development Economics*, 49(1), 137-150.

Kugler, M. 2006. Spillovers from foreign direct investment: within or between industries? *Journal of Development Economics*, 80(2), 444-477.

Lall, S. 1980. Vertical interfirm linkages in LDCs: An empirical study. *Oxford Bulletin of Economics and Statistics*, 42(3), 203-226.

Lara, A.A. and Carrillo, J. 2003. Globalización tecnológica y coordinación intra-empresarial en el sector automovilístico: el caso de Delphi-Mexico. *Comercio Exterior*, 53, 604-616.

Levine, R. and Renelt, D. 1992. A sensitivity analysis of cross-country growth regressions, in *Economic growth in the long run: A history of empirical evidence, vol. 2*, edited by Ark, B. Edward Elgar: Cheltenham, UK.

Lipsey, R.E. 2004. Home- and host country effects of foreign direct investment. In Baldwin, R.E. and Winters, A.L. (eds) *Challenges to globalisation*. Chicago: Chicago University Press.

Lipsey, R.E. and Sjöholm, F. 2004. FDI and wage spillovers in Indonesian manufacturing. *Review of World Economics*, 140(2), 287-310.

Lipsey, R.E. and Sjöholm, F. 2005. The impact of inward FDI on host countries: Why such different answers? in *Does foreign direct investment promote development?* edited by Moran, T.H., Graham, E.M. and Blomström, M. Washington, DC: Institute for International Economics.

Liu, Z. 2001. Foreign direct investment and technology spillovers: Evidence from China. *Journal of Comparative Economics*, 30(3), 579-602.

Liu, Z. 2008. Foreign direct investment and technology spillovers. *Journal of Development Economics*, 85(1-2), 176-193.

Lommel, van E., Brabander, B. and de Liebaers, D. 1977. Industrial concentration in Belgium: Empirical comparisons of alternative seller concentration measures. *Journal of Industrial Economics*, 26(1), 1-20.

Loser, C. and Kalter, E. 1992. Mexico: The strategy to achieve sustained economic growth. *IMF Occasional Papers*, no. 992. Washington, DC: International Monetary Fund.

Love, J.H. and Lage-Hidalgo, F. 2000. Analyzing the determinants of US direct investment in Mexico. *Applied Economics*, 32(10), 1259-1267.

Lutz, S. and Talavera, O. 2004. Do Ukrainian firms benefit from FDI? *Economics of Planning*, 37(2), 77-98.

Magrini, S. 2004. Regional (di)convergence. In, Henderson, V.J. and Thisse, J.F. (eds) *Handbook of urban and regional economics, vol. 4*. Amsterdam: North Holland.

Marshall, A. 1890. *The principles of economics*. London: Macmillan.

Martin, R. 1999. The new 'geographical turn' in economics: Some critical reflections. *Cambridge Journal of Economics*, 23, 63-91.

Martinez-Solano, L.E. and Phelps, N.A. 2003. The technological activities of EU MNEs in Mexico. *International Planning Studies*, 8(1), 53-75.

McFadden, D. 1974. Conditional logit analysis of qualitative choice behaviour. In: Zarembka, P. (ed.) *Frontiers in Econometrics*. New York: New York Academic Press.

Mishan, E.J. 1971. The post-war literature on externalities: An interpretative essay. *Journal of Economic Literature*, 9(1), 1-28.

Mollick, A.V., Duran, R.R. and Silva-Ochoa, E. 2006. Infrastructure and FDI inflows in Mexico: A panel data approach. *Global Economy Journal*, 6, Article 6.

Monfort, Ph. and Nicolini, R. 2000. Regional convergence and international integration, *Journal of Urban Economics*, 48(2), 286-306.

Moomaw, R.L. 1981. Productivity and city size: A critique of the evidence. *Quarterly Journal of Economics*, 96(4), 675-688.

Moran, T.H. 2005. How does FDI affect host country development? Using industry case studies to make reliable generalizations. p. 281-314. In, Moran, T.H., Graham, E.M. and Blomström, N. (eds) *Does foreign direct investment promote development?* Washington, DC: Institute for International Economics.

Moreno, R., Paci, R. And Usai, S. 2005. Spatial spillovers and innovation activity in European regions. *Environment and Planning A*, 37(1), 1793-1812.

Moretti, E. 2004. Human capital externalities in cities. In, Henderson, V.J. and Thisse, J.F. (eds) *Handbook of urban and regional economics vol. 4*. Amsterdam: North Holland.

Nair-Reicher, U. and Weinhold, D. 2001. Causality tests for cross-country panels: A new look at FDI and economic growth in developing countries. *Oxford Bulletin for Economics and Statistics*, 63(2), 153-172.

OECD 2002. *OECD Economic Outlook*, no. 37. Paris: Organisation for Economic Co-operation and Development.

OECD 2007. *Economic Survey of Mexico 2007: Maximizing the gains from integration in the world economy*. Paris: Organisation for Economic Co-operation and Development.

Ottaviano, G.I.P. and Puga, D. 1998. Agglomeration in the world economy: A survey of the 'new economic geography'. *The World Economy*, 21(6), 707-731.

Owell, F.A. 1977. *Measuring inequality*. LSE Handbook in Economics Series. London: Prentice-Hall-Harvester Wheatsheaf.

Pacheco-Lopez, P. 2005. Foreign direct investment, exports and imports in Mexico. *World Economy*, 28(8), 1157-1172.

Padilla-Pérez, R. 2008. A regional approach to study technology transfer through foreign direct investment: The electronics industry in two Mexican regions. *Research Policy*, 73(5), 849-860.

Papandreou, A.A. 1994. *Externality and institutions*. Oxford: Clarendon Press.

Parr, J.B. 2002a. Agglomeration economies: Ambiguities and confusions. *Environment and Planning A*, 34(4), 717-731.

Parr, J.B. 2002b. Missing elements in the analysis of agglomeration economies. *International Regional Science Review*, 25(2), 151-168.

Perez, T. 1998. *Multinational enterprises and technological spillovers*. Amsterdam: Harwood.

Porter, M.E. 1998. *The competitive advantage of nations*. New York: Free Press.

Potter, J., Moore, B. and Spires, R. 2002. The wider effects of inward foreign direct investment in manufacturing on UK industry. *Journal of Economic Geography*, 2(2), 279-310.

Quigley, J.M. 1998. Urban diversity and economic growth. *Journal of Economic Perspectives*, 12(2), 127-138.

Ramirez, M.D. 2000. Foreign direct investment in Mexico: A cointegration analysis. *The Journal of Development Studies*, 37(1), 138-162.

Ramirez, M.D. 2002. Foreign direct investment in Mexico during the 1990s: An empirical assessment. *Eastern Economic Journal*, 28(3), 409-423.

Ramirez, M.D. 2003. Mexico under NAFTA: A critical assessment. *The Quarterly Review of Economics and Finance*, 43(5), 863-892.

Ramirez, M.D. 2006. Is foreign direct investment beneficial for Mexico? An empirical analysis. *World Development*, 34(5), 802-817.

Rey, S.J. and Janikas, M.V. 2005. Regional convergence, inequality and space. *Journal of Economic Geography*, 5(2), 155-176.

Reynolds, C.W. 1970. *The Mexican economy: Twentieth-century structure and growth.* New Haven en London: Yale University Press.

Rigby, D.L. and Essletzbichler, J. 1997. Evolution, Process Variety and Regional Trajectories of Technological Change in US Manufacturing. *Economic Geography*, 73(3), p. 269-284.

Rigby, D.L. and Essletzbichler, J. 2000. Impacts of Industry Mix, Technological Change, Selection and Plant Entry/Exit on Regional Productivity Growth. *Regional Studies*, 34(4), 333-342.

Rodríguez-Oreggia, E. 2005. Regional disparities and determinants of growth in Mexico. *Annals of Regional Science*, 39(2), 207-220.

Rodríguez-Oreggia, E. 2007. Winners and losers of regional growth in Mexico and their dynamics. *Investigación Económica*, LXVI, 43-61.

Rodríguez-Pose, A. 1998. *The dynamics of regional growth in Europe: Social and political factors.* Oxford: Clarendon Press.

Rodríguez-Pose, A. and Sánchez-Reaza, J. 2002. The impact of trade liberalisation on regional disparities in Mexico. *Growth and Change*, 33(1), 72-90.

Rodríguez-Pose, A. and Sánchez-Reaza, J. 2005. Economic polarization through trade: trade liberalization and regional growth in Mexico. 237-259. In, Kanbur, R. and Venables, A.J. (eds) *Spatial Inequality and Development* Leiderdorps Kamerkoort.

Rodríguez-Pose, A. and Crescenzi, R. 2008. Research and development, spillovers, innovation systems, and the genesis of regional growth in Europe. *Regional Studies*, 42(1), 51-67.

Rodrik, D. 1992. Closing the productivity gap: Does trade liberalisation really help? In, Helleiner, G. (ed.) *Trade policy, industrialisation and development: New perspectives.* Oxford: Clarendon.

Rodrik, D. 1999. *The new global economy and developing countries: Making openness work.* Washington: Johns Hopkins University Press for the Overseas Development Council.

Romer, P.M. 1994. The origins of endogenous growth. *Journal of Economic Perspectives*, 8(1), 3-22.

Romero Kolbeck, G. and Urquidi, V. 1952. *La extensión Fiscal en el Distrito Federal como Instrumento de Atracción de Industries.* Departamento del Distrito Federal, México.

Rosenthal, S. and Strange, W. 2004. Evidence of the nature and sources of agglomeration economies, in *Handbook of urban and regional economics, vol. 4*, edited by Henderson, V.J. and Thisse, J.F. Amsterdam: North Holland.

Ruane, F. and Ügur. A. 2005. Foreign direct investment and productivity spillovers in Irish manufacturing industry: Evidence from plant level panel data. *International Journal of the Economics of Business*, 12(1), 53-65.

Sabirianova, S.P., Svejnar, J. and Terell, K. 2004. Distance to the efficiency frontier and FDI spillovers. *CEPR Discussion Paper Series*, no. 4723. London: Centre for Economic Policy Research.

Saggi, K. 2002. Trade, foreign direct investment, and international technology transfer: A survey. *World Bank Research Observer*, 17(2), 191-235.

Sala-i-Martin, X. 1997. I just ran two million regressions. *The American Economic Review*, 87(2), 178-183.

Sargent, J. and Matthews, L. 2004. What happens when relative costs increase in export processing zones? Technology, regional production networks and Mexico's maquiladoras. *World Development*, 32(12), 2015-2030.

Schoors, K. and van der Tol, B. 2002. Foreign direct investment spillovers within and between sectors: Evidence from Hungarian data. *Ghent University Working Paper Series*. Ghent University.

Scitovsky, T. 1954. Two concepts of external economies. *The Journal of Political Economy*, 62, 143-151.

Sgard, J. 2001. Direct foreign investment and productivity growth in Hungarian firms, 1992-1999. *Working Paper Series CEPII*, no. 19. Paris: Centre d'Études Prospectives y d'Information Internationales.

Shaiken, H. 1990. *Mexico in the global economy: High technology and work organization in export industries*. San Diego, CA: Centre for US-Mexican Studies, UCSD.

Silvers, A.L. 2000. Limited linkages, demand shifts and the transboundary transmission of regional growth. *Regional Studies*, 34(3), 239-251.

Sinani, E. and Meyer, K. 2004. Spillovers of technology transfer from FDI: the case of Estonia. *Journal of Comparative Economics*, 32(3), 445-466.

Sjöholm, F. 1999. Productivity growth in Indonesia: The role of regional characteristics and foreign investment. *Economic Development and Cultural Change*, 47(3), 559-584.

Sklair, L. 1993. *Assembling for development: The maquiladora industry in Mexico and the United States*. San Diego, CA: Centre for US-Mexican Studies, UCSD.

Sloboda, B. and Yao, V. 2008. Interstate spillovers of private capital and public spending. *Annals of Regional Science*, 42(3), 505-518.

Smarzynska, B.K. 2002. Does foreign direct investment increase the productivity of domestic firms? In search of spillovers through backward linkages. *World Bank Policy Research Working Paper Series*, no. 2923. Washington, DC: World Bank.

South, R. 1990. Transnational 'maquiladora' location. *Annals of the Association of American Geographers*, 80(4), 549-570.

South, R. 2006. Spatial variations in Mexican maquiladora closure. *Urban Geography*, 27(8), 734-756.

Staiger, D. and Stock, J.H. 1998. Instrumental variable regression with weak instruments. *Econometrica*, 65(3), 557-586.

Svejnar, J., Gorodnichenko, Y. and Terell, K. 2007. When does FDI have positive spillovers? Evidence from 17 emerging market economies. *Institute for the Study of Labour Discussion Paper Series*, no. 3079. Bonn: Institute for the Study of Labour (IZA).

Taki, S. 2005. Productivity spillovers and characteristics of foreign multinational plants in Indonesian manufacturing. *Journal of Development Economics*, 76(2), 521-542.

Taylor, J. 1993. An analysis of the factors determining the geographical distribution of Japanese manufacturing investment in the UK, 1984-1991. *Urban Studies*, 30(7), 209-224.

The Economist 1992. *It's happening in Monterrey*. vol. 323, issue 7765, p. 50.

Thomas, D.E. and Grosse, G. 2001. Country-of-origin determinants of foreign direct investment in an emerging market: The case of Mexico. *Journal of International Management*, 7(1), 59-79.

Torlak, E. 2004. Foreign direct investment, technology transfer and productivity growth in transition countries: Empirical evidence from panel data. *Georg-August-Universitat Göttingen Discussion Paper*, no. 26. Göttingen: Göttingen University.

Tybout, J.R. and Westbrook, M.D. 1995. Trade liberalization and the dimensions of efficiency change in Mexican manufacturing industries. *Journal of International Economics*, 39(1-2), 373-391.

UNCTAD 1992. *World investment report 1992: Transnational corporations as engines of growth*. Geneva: United Nations Conference on Tariffs and Trade.

UNCTAD 2001. *World investment report 2001: Promoting linkages*. Geneva: United Nations Conference on Tariffs and Trade.

UNCTAD 2005. *World investment report 2005: Transnational corporations and the internationalization of R&D*. Geneva: United Nations Conference on Tariffs and Trade.

UNCTAD 2007. *World investment report 2007: Transnational corporations, extractive industries and development*. Geneva: United Nations Conference on Tariffs and Trade.

UNCTC 1992. *Foreign direct investment and industrial restructuring in Mexico*. *UNCTC Current Studies Series A*, no. 18. New York: United Nations.

Vellinga, M. 1995. The Monterrey industry and the North-American market: Past and present dynamics. *Journal of Borderland Studies*, X, 45-68.

Vellinga, M. 2000. Economic internationalisation and regional response: The case of north east Mexico. *Tijdschrift voor Sociale en Economische Geografie*, 91(3), 293-307.

Venables, A.J. and Barba Navaretti, G. 2005. *Multinational firms in the world economy*. Oxfordshire, UK: Princeton University Press.

Viner, J. 1953. Cost curves and supply curves, p. 198-232. In, Stigler, G.J. and Boulding, K.E. (eds) *Readings in price theory*. London: George Allen and Unwin.

Waldkirch, A. 2003. The new regionalism and foreign direct investment: The case of Mexico. *Journal of International Trade and Economic Development*, 12(2), 151-184.

Weis, J. 1992. Trade liberalization in Mexico in the 1980s: Concepts, measures and short-fun effects. *Review of World Economics*, 128(4), 711-726.

Wilson, P. 1992. *Exports and local development: Mexico's new maquiladoras.* Austin: University of Texas Press.

Woodward, D. 1992. Locational determinants of Japanese start-ups in the United States. *Southern Economic Journal*, 58, 690-708.

Wooldridge, J.M. 2002. *Econometric analysis of cross section and panel data.* Cambridge, MA: The MIT Press.

Yamawaki, H. 2006. The location of American and Japanese multinationals in Europe. *International Economics and Economic Policy*, 3(2), 157-173.

Yudaeva, K., Kozlov, K., Mentieva, N. and Ponomaryova, N. 2003. Does foreign ownership matter? The Russian experience. *Economics of Transition*, 11, 383-409.

Zanfei, A. 2000. Transnational firms and the changing organisation of innovative activities. *Cambridge Journal of Economics*, 24(5), 515-542.

Zhang, K.H. 1999. Foreign direct investment and economic growth: Evidence from ten east Asian economies. *Economia Internazionale*, 52, 517-535.

Zhang, K.H. 2001. Does foreign direct investment promote economic growth? Evidence from East Asia and Latin America. *Contemporary Economic Policy*, 19(2), 175-185.

Zukowska-Gagelmann, K. 2000. Productivity spillovers from foreign direct investment in Poland. *Economic Systems*, 24(3), 232-256.

Data-sources

INEGI (1989) *Censos Económicos 1988.* Aguascalientes: Instituto Nacional de Estadística y Geografía.

INEGI (1991) *XI Censo General de Población y Vivienda 1990.* Aguascalientes: Instituto Nacional de Estadística y Geografía.

INEGI (1994) *Censos Económicos 1993.* Aguascalientes: Instituto Nacional de Estadística y Geografía.

INEGI (1996) *Conteo de Población y Vivienda 1995.* Aguascalientes: Instituto Nacional de Estadística y Geografía.

INEGI (1999) *Censos Económicos 1998.* Aguascalientes: Instituto Nacional de Estadística y Geografía.

INEGI (2001) *XII Censo General de Población y Vivienda 2000.* Aguascalientes: Instituto Nacional de Estadística y Geografía.

INEGI (2004) *Censos Económicos 2003.* Aguascalientes: Instituto Nacional de Estadística y Geografía.

INEGI (various years) *Anuario de Estadísticas Estatal*. Aguascalientes: Instituto Nacional de Estadística y Geografía.

INEGI (various years) *Sistema de Cuentas Nacionales de México. La Producción, Salarios, Empleo y Productividad de la Industria Maquiladora de Exportación.* Aguascalientes: Instituto Nacional de Estadística y Geografía.

Index